WINGS
OF
WAR

WINGS
OF
WAR

The World War II Fighter Plane
That Saved the Allies and
the Believers Who Made It Fly

David Fairbank White
and
Margaret Stanback White

CALIBER

DUTTON CALIBER

An imprint of Penguin Random House LLC
penguinrandomhouse.com

Maps by Chris Erichsen Cartography

LIBRARY OF CONGRESS CATALOGING-IN-PUBLICATION DATA

Names: White, David Fairbank, 1951– author. | White, Margaret Stanback, author.
Title: Wings of war: the World War II fighter plane that saved the Allies and the believers who made it fly / David Fairbank White, and Margaret Stanback White.
Description: [New York]: Caliber, [2022] | Includes bibliographical references and index.
Identifiers: LCCN 2022026360 (print) | LCCN 2022026361 (ebook) |
ISBN 9781524746322 (hardcover) | ISBN 9781524746346 (ebook)
Subjects: LCSH: World War, 1939–1945—Aerial operations, American. | World War, 1939–1945—Aerial operations, British. | Mustang (Fighter plane)—History. | United States. Army Air Forces—History. | Great Britain. Royal Air Force—History. | World War, 1939–1945—Campaigns—Europe. | Fighter pilots—United States—Biography. | Fighter pilots—Great Britain—Biography.
Classification: LCC D785 .W44 2022 (print) | LCC D785 (ebook) |
DDC 940.54/4—dc23/eng/20220608
LC record available at https://lccn.loc.gov/2022026360
LC ebook record available at https://lccn.loc.gov/2022026361

Printed in the United States of America
1st Printing

For Thomas M. Stanback, Jr.,
20th Fighter Group, Eighth Air Force,
U.S. Army Air Forces,
1943–1945

It was cold at the airport and dark. . . . I walked round my ship, stroking her wings with the back of my hand in a caress that I believe was love. Eight thousand miles I had flown in her, and her engines had not skipped a beat; not a bolt in her had loosened. This was the marvel that was to save our lives . . . by refusing to be ground to powder on meeting the upsurging earth.

—Antoine de Saint-Exupéry, *Wind, Sand and Stars,* 1939

Contents

CONTENTS

PART THREE: HAWK OF HEAVEN

PROLOGUE

The Landmarks of a Nightmare

These were the big bombers, massive 18-ton B-17 Flying Fortresses with four Cyclone radial engines, two to each wing, capable of almost 290 miles per hour, and now they were coming in over Belgium, high above the clouds, over Diest and Antwerp. They were 75 feet long, longer than a highway truck trailer; had a majestic wingspan of 103 feet, seven inches, nearly the length of a baseball diamond; and carried a crew of ten—pilot, copilot, navigator, bombardier and six others. Twelve .50-caliber-machine-gun mounts bristled from the fuselage and more than 200 of these heavy bombers, covering 10, 13, 15 miles of contrail and sky, were thundering in on approach and closing. They carried nine tons of bombs and roared on, not diverting or slowing or wavering, no matter what they met. And that autumn they met havoc. It was October 14, 1943. Their target was Schweinfurt, in the interior of Germany, a plant for ball bearings used in aircraft production. They homed in, howling on at 35,000 feet, seven miles above the clouds in the radiance of the sun, moving in like the front of a typhoon.

Their escorts were stubby Republic P-47 Thunderbolts, known as "Jugs," short-range fighters with a limited endurance of only 375 miles. The Jugs would guard the two columns across the fields of Belgium to the German border, but they could not shepherd the B-17s into the interior of Germany. Over Aachen, the Thunderbolts reached the limit of their fuel. They waggled their wings in farewell and peeled off toward home, leaving their charges unprotected. Now the bombers were alone over Mainz, near Frankfurt, all configured in combat "boxes" to form an airborne fortress in the wide blue dome of the sky.

The German fighters struck then. They came in waves, two, three, four at a time, reptilian Focke-Wulf Fw 190s, the latest and fastest German design. Major General Adolf Galland, commander of German fighter forces, would later reveal that the Luftwaffe had marshaled virtually its entire strength to defend Schweinfurt on that day. The result was nothing short of slaughter.

Focke-Wulfs veered and swooped through the formation of B-17s. Larger, rocket-firing Junkers Ju 88s sent missiles into the U.S. squadrons that exploded in bursts four times larger than ground fire. Some 300 German fighters, 40 fighter-bombers and other aircraft joined the fracas, diving, plunging, sweeping across miles of sky. Now the fighting was intense. Fortresses were struck, erupting into flames, smoke gusting from their engines, yellow blazes filling the sky, black puffs of flak appearing all around. One American bomber crewman recalled "the landmarks of a nightmare" as he clocked the Fortresses falling, cartwheeling down the sky, and tried to count the strings of parachutes opening as men bailed out. Ten parachutes meant an entire crew had escaped a burning plane, but there were seldom ten.

Of 229 Fortresses that reached the target that day over Schweinfurt, 60 were shot down and 17 more were lost on the way home, a total of 77 sky arks gone for a loss rate of 26 percent. A total of 642 men were lost out of a force of 2,900, or 22 percent killed in action. These were unsustainable losses. At this rate, the Allied air forces would go down in a ghastly war of attrition, and the Luftwaffe would achieve mastery of the air. October 14 came to be known as "Black Thursday."

The October Schweinfurt raid capped four months of carnage in the air. That fall at sea, the Battle of the Atlantic was desperately fought in heavy swells and moaning winds. In the frigid east, on land, the Russians were locked in a death struggle with German armies in Operation Barbarossa. Now the air war, too, was failing, with catastrophic missions over Bremen, Anklam, Münster, Stuttgart, Marienberg, and once before over Schweinfurt. With the bombers rode Allied fortunes, and in October 1943 the bombers were being shot down by the hundreds. The Allied air campaign was reaching the floor of a very deep pit.

Aircrews were grounded. Some men talked of mutiny. American generals were out of answers for guarding the bombers, and the outcome of the whole war seemed to stretch for one suspended moment and almost come apart.

It was at this bleak moment in the fall of 1943 that a new American weapon would appear, ready to charge like a cavalry brigade into the European altitudes. It was the swiftest, most nimble aircraft the U.S. Army Air Forces had yet developed, a futuristic plane so fast it could outfly any German fighter in the skies over Europe. It could reach speeds of over 400 miles per hour, and it could climb 2,300 feet per minute, 38 feet in one second. It was powerfully armed with six .50-caliber machine guns in its wings and racks for bombs. Above all, it would eventually have a range of 2,080 miles, long enough to accompany the bombers to any target inside Germany and back to England.

This was the North American P-51B Mustang—a pursuit plane crash-designed in 102 days by a visionary designer with an almost narcotic passion for flight from boyhood. It was the best fighter plane of World War II, the greatest piston engine fighter ever made. But the P-51 was not just fast and acrobatic. It was trailblazing, with pioneering innovations from engine to wingtip. It incorporated breakthroughs—from its thrust, to its specialized streamlining, to its balance, to its massive gas tanks—that knocked on the door of the jet age and broke open a new frontier in aviation design.

At this critical hour, when Allied bombers were being picked off like geese over Germany, the P-51 would arrive to change the calculus of failure. In less than six months, virtually single-handed, it would decimate the German Luftwaffe, turn around the war in the skies and open the way for the D-Day invasion. It became the hinge on which all the war in Europe finally swung.

Yet, for all the glory it would attain, the Mustang was for a long time an outcast. From its design in 1940 until 1943, the U.S. Army Air Forces would ignore it as thousands of young airmen went to their deaths. That story, once exposed, would be one of bureaucracy, cronyism, greed and outdated beliefs. It would represent to many the costliest mistake the Allies made during the war.

The Mustang finally began arriving at an American air base in Boxted, near Colchester, England, in November 1943. It had traveled all the way from California, where a small upstart aviation company had designed and built it three years earlier. By early 1944, the Allies would begin flying it to victory. Three men were indispensable in bringing the bird to the skies over Europe. One was a dreamer, spellbound by his obsession with flight, who designed the Mustang. One was a golden American titan, straddling his age in moneyed success, who overcame resistance to the plane. And one was the boldest fighter ace of the war, who first flew the Mustang in combat and lost his heart to it. The story begins with the dreamer.

PART ONE

THE PLANE IN
THE DREAM

Edgar Schmued

CHAPTER 1

Hypnotized

It was high up in the sky, far above him, droning and gliding on, and at first the boy could not tell what it was and kept fixing the spot with a spellbound gaze. But then it continued, and yes, it was an airplane, a Wright Flyer on a course out of Germany, heading toward Russia. This was just a few years after the Wright brothers had first launched their primitive biplane off the beaches of Kitty Hawk, North Carolina, in 1903. But it was a long way from the sands of the Carolinas' Outer Banks.

As the plane cruised above the ancient medieval town of Landsberg, it caught the eye, then the whole attention, then the rapt wonder of twelve-year-old Edgar Schmued. With dark hair and deep-set brown eyes, his neck cocked back and his gaze steady, Edgar was hypnotized by the sight. It was on this day in 1912, as he gazed over the rooftops of Landsberg, nestled on the banks of the Lech River in Bavaria, that Edgar's life was kidnapped by aviation.

He had no interest in schoolwork. Instead, he lived in a private world of daydream and fantasy about machines and how they worked. He imagined how paper was made, how gears engaged and meshed, how spinning machines spun thread. He whiled away his days studying pictures of the latest inventions, making sketches of transmissions, gears, engines. His father, an Austrian dentist, shared Edgar's passion for machines and bought his son every technical book he could find, spending hours explaining how dynamos and electrical circuits worked. But Heinrich Schmued was not always a good provider. With five children besides Edgar—Erwin, Erich, Eugenie, Elfriede and Else—the family sometimes went hungry, and poverty would soon shipwreck Edgar's hopes. Dr. Schmued had no money to send his son to university. When the time came, the most the father could do for the boy was to buy him the textbooks.

Forced to chart his own path, Edgar assembled a makeshift curriculum, setting for himself a rigorous course of reading about the machines that so captivated him. He apprenticed himself to a local engine factory, where he learned the skills of metalworking—milling, lathing, forging by hand. At the end of his two-year stint, he built his own motor. And all the while his spellbound fascination with flight burned on.

Edgar's impromptu education and apprenticeship were unfolding in the contrail of developments that were transforming the globe. The Industrial Revolution had ushered in a new age of steam and steel. The Bessemer process for purifying and producing steel from iron ore in 1856, Thomas Edison's development of cheap electricity for the masses in the 1880s, reciprocating engines, telephones linking distant continents, the exploding artillery shell and more—all would alter the world like a wave rushing across the surface of the sea. This vast metamorphosis was transforming ships with propellers, mail with telegraph, horses with motorcars. And the first forays into manned flight were just capturing the public's imagination.

The spectacle of birds gliding free of the earth, unbounded in the open sky, had mesmerized since ancient times. The fantasy of men soaring through the air had lured inventors and scientists in the earliest fables of Greek myth, when Icarus was said to have flown too close to the sun, melting his wings of wax and feathers. And now the beginnings of manned flight, a brave new science that would soon open an uncharted wilderness around the globe, were quickly morphing from one form to the next toward that ethereal goal of soaring.

Aeronautics was so young that its few practitioners were practically celebrities. The very first manned ascent—by the French Montgolfier brothers in their hot-air balloon in 1783—and Sir George Cayley's first manned flight in 1853 opened the atmospheres to the wing, and these milestones still stirred the imagination in the early 1900s when Edgar was a boy.

Cayley, often known as the father of aviation, was an English aristocrat growing up on his father's estate in Yorkshire when he became fascinated by stories of early balloon flights. From his daydreams of lighter-than-air rides in hot-air balloons, he began to think about how to keep aloft a machine that was heavier than air. By observing birds in flight, and systematically experimenting with different shapes of wings, Cayley was able to describe the problems of "lift" and "drag" as key to soaring. He realized that the streamlined shapes of creatures in nature—birds moving through the air, fish moving through water—minimized drag and would be ideal for a flying machine. In one dramatic inquiry, he sliced open a trout crosswise into sections and studied the streamlining of the body of the fish. By 1809 he already had the basic shape of an airplane's wing.

The German Otto Lilienthal, too, was mesmerized by birds. In the late 1800s Lilienthal also believed bird flight and anatomy held the formula for heavier-than-air flight. Known as the Birdman, Lilienthal concluded that "cambered" or curved wings were the best shape to produce lift. His "ornithopters" were flimsy cotton-and-bamboo structures that he flew with great showmanship by flexing and twisting the birdlike wings. The Wright brothers would soon develop wings that flexed and twisted, too—the forerunners of modern flaps and ailerons.

Lilienthal soon went on to build gliders, which he piloted by throwing himself around to compensate for shifting and often treacherous air currents. He would make more than 2,000 flights in his various contraptions, the longest 1,150 feet. The last, when his glider caught a gust of wind, would kill him in 1896.

And so manned flight staggered and lurched ahead, but powered flight was still a fantasy beyond reach. The late 1800s brought steam, but steam-powered engines were heavy and bulky, useless for soaring. The Frenchman Félix du Temple de la Croix had briefly lifted off in a steam-powered "air car" in 1874—but the vessel had fallen back to earth. In 1896, Samuel Pierpont Langley, the secretary of the Smithsonian Institution, experimented with steam-powered "aerodromes." When those failed, he began to experiment with gasoline engines for his machines. Percy Sinclair Pilcher, an Englishman, was the first to fit an internal combustion car engine to his "bat glider" in 1899. Pilcher was killed in a crash during a test flight, but he had found the propulsion for all future airplanes: gasoline.

As the century waned, no one asked if men would fly, only when.

And then, in 1903, the Wright brothers made their historic flight from the bare, windswept beach at Kitty Hawk, North Carolina. On the morning of December 17, Orville Wright took his position on the bottom wing of the flimsy wood-and-cloth biplane the brothers had made in their bicycle shop in Dayton, Ohio, facing into the chilly winter wind. Behind him two propellers were connected to a 12-horsepower engine by chains and pulleys. As Wilbur ran alongside, the machine lifted ten feet into the air and flew for twelve seconds. Orville made three more flights in the fragile biplane that day; on the last sortie, the craft actually flew for 59 seconds, covering 852 feet. Manned, powered flight on a piloted, heavier-than-air machine was born.

In 1904, the Wright brothers stayed aloft for five minutes. In 1905, they made a flight of 38 minutes, covering 20 miles; in 1908, after a Paris demonstration of their machine, Wilbur made a flight of two hours and 20 minutes and set a new altitude record, climbing to 360 feet. He carried, on various trips, a total of 60 passengers— hinting at the immense possibilities for airline passenger travel. By 1910, the Wrights were among the most famous men in the world, their achievements galvanizing the public.

Alberto Santos-Dumont, a Brazilian living in Paris, was first in flight in Europe, his box-kite aircraft, the 14-*bis*, traveling 722 feet in 21 seconds in 1906. He was at once a French national hero. And then, in the still air of a summer's day in 1909, the great French aviator Louis Blériot became the first man to fly across the English Channel, staying aloft for 36 minutes at an average 40 miles an hour, and claiming a fat £1,000 prize offered by the British *Daily Mail.*

As Edgar grew up, the magic of flight hijacked the boy's imagination. A voracious reader, he studied every text on aviation he could find. One of his favorite books was *The Resistance of the Air and Aviation Experiments* by Gustave Eiffel—the famed Frenchman who had built the Eiffel Tower in Paris for the World's Fair in 1889. Edgar saw an early copy in 1911.

Eiffel was looking into a horizon as wide as the wind and altitude, into the barely born field of aeronautics. Fascinated by the effects of wind on the structures he had built, Eiffel was among the first to explore drag and lift, Cayley's twin riddles of flight. He experimented by dropping objects from railway bridges, viaducts, even the Eiffel Tower itself. He built the world's first wind tunnel to test the properties of "airfoils," or wings. Observing the Wright brothers and the Frenchman Blériot, Eiffel showed for the first time that lift is the result of reduced air pressure *above* the wing, not of increased air pressure *below* it. Lift made flight possible—not flapping, riding on the air, gliding or floating. This was a stunning new concept in 1911. Eiffel was sketching the first laws of aerodynamics, and the young Edgar drank them down like a heady draft of gin. He began thinking about how to reduce pressure above the wings of an airplane. The question would stay in his mind for many years.

Aviation soared into the open sky. Soon it would become a big business. Two French brothers, Gabriel and Charles Voisin, would start up the world's first airplane factory at Billancourt on the outskirts of Paris in 1906. By 1911, no fewer than eight aircraft makers had sprung up in France. Before long the Wrights were also manufacturing their airplane in Europe. In 1912, Vsevolod Abramovich, a 22-year-old Russian test pilot for the Wright brothers' subsidiary in Germany, flew one of the early Wright Flyers to a competition in St. Petersburg.

It was this plane and this pilot, on this day, that the boy Edgar Schmued saw overhead. Abramovich could not have known he had a vicarious passenger traveling with him through the clouds.

World War I broke out in folly and waste that would cost 8.5 million lives in pitiless trench warfare and devastating artillery barrages. It also brought the horrors of modern combat to what had been the rowdy carnival of aviation.

As the war began, naval fleets dominated. The airplane was no more than a toy, a delicate wood-and-wire experiment piloted by men in goggles and leather helmets who were largely daredevils and thrill seekers. The first planes carried no guns—no weapons of any kind. They were reconnaissance aircraft used to pinpoint targets for artillery and report on enemy-troop concentrations. But flying soon lost its innocence. First, pilots began taking pistols and rifles up into the air with them; then machine guns were mounted on fighters. The key leap forward was the invention of the "interrupter," which synchronized the timing of machine guns so they could shoot through the gaps between whirling propeller blades. This 1915 breakthrough was made by Anton Fokker, a charismatic Dutchman who would soon play a pivotal role in Edgar Schmued's life.

Fokker made aerial gunnery possible, and the airplane charged into war.

New planes invaded the skies over Europe—the British Spad XIII and Sopwith Camel, the French Nieuport Type 17 and Morane-Saulnier Model N, the German Albatros D.III and Fokker Dr.I, all fighter planes and all armed. By the end of the war, bombers would appear over Allied cities, and the Germans would use giant zeppelin airships to raid London—but the fighters still took center stage. The spectacular gladiatorial exploits of the famed fighter aces were the most celebrated actions of World War I.

Highest scoring was the great German ace Baron Manfred von Richthofen, the "Red Baron," who led the tally with 80 enemy planes shot down. A Prussian nobleman who had ridden in the cavalry and was an avid hunter, von Richthofen was known for his caution, consummate discipline and studied dislike of the French. His fighter wing became known as von Richthofen's "Flying Circus."

The top American gunslinger was Eddie Rickenbacker. As a race car driver in his youth, he had four times driven in the Indianapolis 500 and set a land speed record of 134 miles per hour in a Blitzen Benz. Rickenbacker shot down 26 planes in only two months of combat, a monthly average far higher than von Richthofen's. These duels were fast, acrobatic and deadly. In one action, Rickenbacker ran into six German fighters over Étain, France. He closed on them, climbed swiftly into the sun unnoticed, got behind the German Fokkers and charged, attacking the nearest one. He fired one fast burst at the Fokker, scoring a hit and sending it down. Then Rickenbacker dashed toward an LVG, another German plane, homing in. With some aerial maneuvers he hit the LVG, too, and watched it fall from the sky and crash in a trail of black smoke. Rickenbacker had bagged two Germans in one day.

Edgar escaped the destruction of World War I. His obsession with flight became his guardian angel, sparing him from the battlefield. Assigned as a mechanic in the Austro-Hungarian Air Service, he spent the war repairing and maintaining the new military biplanes and triplanes—Aviatics, Hansa-Brandenburgs, Lloyds, Ufags and Fokkers. When the war ended in 1918 and peace settled over Europe, Schmued, now 18, went home to Landsberg.

There in his father's house, with the wartime airplanes he had worked on still vivid in his mind, he set out to build his first flyer from scratch. The original Schmued avian was a biplane; Edgar built it in his father's study. Dr. Schmued even bought a small three-cylinder Belgian engine for Edgar's first machine. When it was finished, the dentist, an amateur inventor himself, knocked down one wall of his study so he and his son could roll the wings and fuselage into the light of day. But before it could be flown, politics doomed the homemade plane. The terms of the Versailles Treaty had required Germany to deliver to the Allies "any aeronautical material including aircraft engines." The Allied Control Commission confiscated Edgar's creation. Thus ended Schmued's first attempt to build an aircraft—but his sister Else remembered that he never stopped drawing them.

Edgar's fascination with aviation would soon pull him much farther from Landsberg than he might ever have imagined. Within a few years this impoverished boy, with his makeshift training and limited horizons, would leave Germany to chase his dream of designing airplanes.

CHAPTER 2

Flugzeug Konstrukteur

Weimar Germany after World War I was no place for a young man with dreams. It was a desolate, turbulent hinterland of chaos and despair. Sky-high inflation wracked the economy. Unemployment was rampant. There were food shortages and widespread hunger. German workers, sapped by the gruesome conflict, were weary and hopeless. When Germany defaulted on its war reparations, France and Belgium occupied the Ruhr Valley, the heart of Germany's industrial region. Eight months of strikes followed. Postwar Germany was a cold, gray field swept by deprivation and roiled by violent factions.

By 1921, in this underworld of ruin, Edgar was working near Hamburg, designing auto parts. He had married a divorced woman nine years his senior, and they had a young son. Within a year he had patented five of his ideas for fuel devices. But massive postwar inflation made backers for his inventions hard to come by, and airplanes were still on his mind. In 1925, he left Germany forever, setting out for Brazil, where two of his brothers had immigrated to the large German community in São Paulo. He was looking for investors for an airplane venture, hoping for a lucky break.

Edgar's gamble was not as far-fetched as it may have seemed. By the 1920s, the aviation craze had taken over Brazil. The famous Alberto Santos-Dumont, first to fly in Europe, was a Brazilian; pride in their country's native son fueled a national mania for flying. Newspapers covered the attempts to fly the Atlantic from Europe to South America on their front pages. The potential of the airplane to shrink the distances of this vast country stood out clearly. When Edgar arrived in 1925, the commercial possibilities for aviation in Brazil seemed boundless. But at first fortune did not favor Schmued.

His stabs at forming an airplane company failed. He could not interest anyone in Brazil in his venture. "I was living with my brother," he said later. "I had to find some way to make a living." With typical grit and determination, Edgar soon managed to right himself. In quick succession, he proceeded to get a job as an auto mechanic for General Motors and meet a GM executive who was so impressed by his ideas for servicing cars that the man assigned him to head up field operations. His boss encouraged him to move to the U.S. "so I could really utilize my ideas and use them properly," Schmued remembered.

In America, manned flight was changing as fast as a scirocco wind. The 1920s were the heady adolescence of flying, when barnstormers, stuntmen, racing pilots and aerial circuses delighted and thrilled small-town America. Out-of-work pilots from the First World War found work as "gypsy fliers," touring rural towns with air shows for local crowds, giving flights for a dollar a ride. Their repertoire was daring; aerial acrobats did wing walking, crossed from one airplane to another in midflight and performed breathtaking "death falls," saved at the last minute by parachutes. These dare-devil stunts electrified the public and drew huge crowds.

The air was open, thrilling in a way that ships, cars, trucks and trains could never be. Soon young entrepreneurs saw a vast, rich frontier before them in aviation. In Europe, German and French interests began to cast lines of service far across the globe. Deutsche Luft-Reederei, an early German enterprise, began operating in 1919, as the Versailles Treaty permitted commercial, though not military, aviation; by 1926 these German pioneers, now Deutsche Luft Hansa, were extending their systems north to Scandinavia, east to the Soviet Union through Poland and south to the Balkans and the Mediterranean. The bold and venturesome pilots of Aéro-postale, the French airmail line known as *La Ligne*, extended the reach of flight to France's far-flung colonies, bushwhacking routes down through North Africa, then transatlantic to South America and overland to the Andes and Chile.

The stories of the exploits of the French airmail pilots read like novels: Henri Guillaumet, who crashed his plane in the Andes Mountains in bad weather while crossing them for the ninety-second time, and walked for a week over three mountain passes to safety; Jean Mermoz, who after a forced landing in the Andes once prevented his airplane from rolling into a ravine by pure physical force; and the renowned author-flier Antoine de Saint-Exupéry, bard of early aviation, who survived a crash in the Libyan desert and was rescued near death on his third day without water by a Bedouin on a camel. He would later write about the experience of flying an airplane:

> There is a particular flavor about the tiny cabin in which, still only half awake, you stow away your thermos flasks and odd parts and overnight bag; in the fuel tanks heavy

with power; and best of all, forward, in the magical instruments set like jewels in their panel and glimmering like a constellation in the dark of the night. The mineral glow of the artificial horizon [display], these stethoscopes designed to take the heart-beat of the heavens, are things a pilot loves. The cabin of a plane is a world unto itself, and to the pilot it is home. . . .

In America, too, such daredevil pilots were recruited by the U.S. Post Office to fly the mail coast-to-coast, overcoming daunting obstacles to navigate over mountains and deserts without maps or reliable compasses and with only primitive instruments. These pistol-packing aviators found their way across the country by following railroad tracks or looking for familiar landmarks like a church steeple or a water tower. Fog and storms posed terrible risks. One pilot who had begun as a barnstormer and then flew the mail from St. Louis to Chicago was Charles Lindbergh. In dense fog and then a snowstorm, he was forced to bail out of his plane and parachute to safety twice within a few months.

As these intrepid pilots extended air routes outward over land, others were conquering a different wilderness: the Atlantic Ocean. In 1919, two Englishmen, John Alcock and his navigator, Arthur Whitten Brown, had broken open the future with the first transatlantic flight, flying from St. John's, Newfoundland, to Galway, Ireland.

Then in 1927, the tall, bashful Lindbergh attempted the first solo transatlantic flight from New York to Paris, a trip that had already cost the lives of six pilots. Overnight, Lindbergh's bravura would shrink the map of the world. Taking off from muddy Roosevelt Field in New York with five sandwiches and two canteens of water as his only provisions, he brought his Ryan NYP *Spirit of St. Louis* through darkness, storm clouds, fog and freezing cold that iced his wings. Flying with only minimal instruments, by the end of his journey he was hallucinating from exhaustion and seeing mirages of land.

On the night of May 21, after 33.5 hours in the air, he dipped into the golden glare of the searchlights at Le Bourget Airport in Paris and landed amid an ecstatic crowd of 150,000 people. His feat announced to the public far and wide that long-distance travel by air had arrived.

Back in the U.S. his reception was no different. Lindbergh made a tour of every state in the union. A ticker-tape parade in New York City drew four million people. In small towns, fans shredded their phone books to throw confetti at local parades and church bells rang in his honor. His likeness appeared everywhere—on statues, wallpaper, tapestries, hooked rugs. "People behaved," his biographer wrote, "as though Lindbergh had walked on water, not flown over it."

Lindbergh's 1927 transatlantic feat would inspire other daring long-distance flights in the 1930s. Women, too, were pushing the envelope: Amy Johnson, a stenographer from Hull, England, made headlines in 1930 when she flew a de Havilland Moth from England to Australia. In 1932, Amelia Earhart became the first woman to fly the Atlantic solo, marking the fifth anniversary of Lindbergh's flight. Beryl Markham, a Kenyan bush pilot, would be the first to fly the Atlantic solo and nonstop, in reverse, from Britain to North America in 1936.

One year later, Earhart took off in her Lockheed Electra with her navigator, Fred Noonan, on a round-the-world tour that would take them out over the void Atlantic Ocean tracts, to Africa and the South Pacific. They were never seen again.

But the most enduring effect of Lindbergh's flight was on commercial aviation. In America, which had lagged behind Schmued's home continent of Europe in aeronautics in the twenties, air transport for both freight and passengers now became a new gold rush, with new routes and new kinds of traffic set to make fortunes for brash entrepreneurs. A timber magnate named William E. Boeing won the overland western U.S. mail route in 1929—and he also made the airplane to fly it. His successful Boeing 80A aircraft carried U.S. airmail, but Boeing also added room for 18 passengers, foreseeing the limitless future of air travel.

Donald Douglas had watched the Wrights fly at Curtiss Field on Long Island in 1908 and witnessed the birth of naval aviation while at the U.S. Naval Academy in Annapolis. In 1920, he moved to California to start an aircraft company with a $40,000 investment and an office in a barbershop. It would soon become the legendary Douglas Aircraft Company and deliver the famed DC-3 airliners, some of the finest piston-engine aircraft ever made. The fierce competition between Boeing and Douglas for the airliner market would propel aircraft design in the U.S. in the 1930s.

Soon a bulging age of new aviation technology brought air travel into the modern world. American pilots became adept at night flying and instrument flying. Radio navigation beacons permitted airliners to fly "on the beam." The first air traffic control tower with radio would be set up in Cleveland in 1930. Stewardesses appeared in 1930, then sleeping accommodations on board. Passengers delighted in the new mode of transportation. These advances made mass-market aviation possible, and commercial airlines multiplied and flourished—Eastern Airlines in 1926; United Airlines, Boeing's brainchild, in 1928; American Airlines, formed from a scatter of 80 smaller companies, in 1930.

Automobile manufacturers began to move into aircraft production to compete with the likes of Boeing and Douglas. Among the first was Schmued's employer, General Motors, which acquired a controlling interest in the storied Fokker Aviation Company in 1929. By the end of the 1920s, Fokker was the largest aircraft manufacturer in the world. Through his lowly job as a car mechanic, Schmued's lifelong dream was beginning to roll into view. He was about to launch himself into aviation.

Schmued applied for a visa from the Austrian consulate and, with only the handmade plane he had made in Landsberg to his credit, proudly described himself on his application as a *"Flugzeug Konstrukteur"*—airplane designer. He soon received U.S. Quota Immigration Visa #1788 from the American vice-consul and embarked for the United States to begin work at Fokker. There, he would finally begin to realize his fantasy of designing airplanes.

Edgar Schmued arrived by ship in New York City, just miles from the Fokker plant in Hasbrouck Heights, New Jersey, in the first weeks of 1930. A photo taken around that time shows an earnest young man with a direct gaze, sporting a well-trimmed mustache and the stiff collar of the era. Like most young immigrants of the day, Schmued was determined to succeed in his new land. He enrolled for 91 classes in English and American literature at the Broadway Evening School for the Foreign Born in nearby Hackensack in order to learn English as quickly as he could. The school noted that his "quality of work was commendable."

Starting at Fokker Aircraft in February 1930, Schmued was joining a firm whose founder was already a legend in aviation. Born in Java to a wealthy Dutch family, Anton Fokker was a daredevil flier who became a wily businessman. During World War I, he designed hundreds of aircraft at his factory in Germany for the

Imperial German Air Service, including the Fokker Eindecker and the fabled Fokker Dr.I. When the Treaty of Versailles forbade Germany to produce military airplanes, Fokker smuggled airframes and engines across the German border to the Netherlands. By 1927, he had established the U.S. arm of his business, the American Aircraft Corporation, near Teterboro Airport in New Jersey. While most planes being built were still biplanes and triplanes, he was boldly beginning to build monoplanes at the Teterboro plant.

Here at Fokker, across the Hudson River, Schmued would finally reach the destiny that he had journeyed toward for 31 years: he would at last design the magical flying machines that had captivated him all his life. It must have seemed to him that he had arrived in Oz, in his own Emerald City.

Fokker ran a tight ship: the workmen, mostly Dutch and German craftsmen he had brought over from Europe, wore white uniforms and were exhorted to be "clean, neat and exact." Edgar rose quickly in the ranks: with Fokker himself still chief engineer, he set up the first "preliminary design" team in the country, a group for developing initial prototypes of a model. And both he and Fokker were busy. Despite the stock market crash the previous year, Fokker Aircraft prospered, milling out its Army XO-27 and XA-7 transports and the F-10 Super Trimotor, flown by leading early airlines including Pan American and TWA. Business boomed.

And then, on March 31, 1931, disaster struck. A Fokker F-10 carrying famed Notre Dame football coach Knute Rockne crashed in a Kansas cornfield, killing Rockne and stunning the nation. Tributes poured in from the likes of President Herbert Hoover and the king of Norway. Fokker's monoplanes consisted of "a welded steel tube, fabric-covered fuselage and wooden wings": the wooden wings were blamed for the crash. The F-10 was grounded and

Fokker's reputation was ruined. He resigned, and the Hasbrouck Heights factory closed. General Motors moved the factory to Dundalk, on the waterfront in Baltimore's inner harbor.

Schmued; his wife, Luisa; and their young son, Rolf, all moved to Dundalk, and at last Edgar seemed to prosper: he bought a red sports car and enrolled in the National Geographic Society. But the plant was little more than a gloomy building with just the skeleton of one F-10 trimotor transport still in the hangar. No more than 75 workers were still in the place. Bankrupt, the company had just one tiny order for a military trainer. Every effort to sell more aircraft had failed. It seemed unlikely that Fokker Aircraft would survive the Depression.

All that would change under a new manager, James H. "Dutch" Kindelberger, who would soon become a prime mover of modern aeronautics. Kindelberger was the son of an iron molder from Wheeling, West Virginia. From high school dropout earning $5 for a six-day week and draftsman for the Army, he had joined Glenn Martin and then Douglas Aircraft in 1925, working on the fabled DC-1 and DC-2 airliners. At 39, he took over the leadership of North American, the holding company for Fokker; he lost no time turning the company around.

Kindelberger, an "extrovert with an extraordinary capacity for hard work," pumped fresh blood into North American. He began by moving the company to California, where the weather was ideally suited to flying. There was a ready labor pool on the West Coast and a fast-growing economy was evolving under the soft sun along the Pacific.

Others had paved the way. Primeval U.S. aircraft manufacturing had begun in the Northeast and around the Great Lakes, close to essential raw materials like iron and coal. But in the 1930s, that began to change. Don Douglas had started out in the West. Allan and Malcolm Loughead (soon to be Lockheed), John Northrop, William Boeing, as well as Reuben Fleet of Consolidated—all soon came, too, for the year-round flying weather and lower heating and real estate costs.

By 1935, Dutch Kindelberger's move was complete. He had leased 20 acres at Mines Field, later to become Los Angeles Municipal Airport and finally LAX, just miles from Santa Monica, where he had worked for Donald Douglas a year earlier. The rent on the property, at the corner of Imperial Highway and Aviation Boulevard in the Los Angeles suburb of Inglewood, was $600 a year. He soon broke ground on a dusty lot near the mission-style buildings of the airport, and by 1935 was moving his employees, many of them the Dutch and German workers Fokker had hired in the late twenties, from Baltimore into his modern white 172,000-square-foot California plant with views of the airport and the Santa Monica mountains.

But Schmued was not among those who had moved west by the fall of 1935. He hesitated to take his family out to the Pacific; Luisa, so recently arrived from Germany, was reluctant to live even farther from her native country. Kindelberger kept the position open, and when it became clear that a new job at Bellanca Aircraft in Delaware held nothing for Edgar, the decision was made. The Schmueds were westward bound. Edgar packed his wife and young son into the family car in November 1935 and set out at last for the promise of California and North America. Then, as they approached Los Angeles, tragedy struck.

On Route 66, just over the California line near the town of Indio, the Schmueds' car was hit head-on by an elderly Kansas couple. Luisa died later in the hospital. Rolf was only cut and bruised, but Edgar sustained a concussion and a broken leg and needed ten stitches to close a torn eye. The accident put him out of commission in the Coachella Valley Hospital for several months. The teenage Rolf was placed with good friends Richard and Liesel Thaiss, another German family recently arrived from Baltimore, who took care of the boy and gave him a bike for Christmas.

When Edgar left the hospital in January 1936, he took a small house at 9600 Redfern Avenue that still stands today—a low, one-story Spanish-style bungalow with a red terra-cotta roof in a neighborhood on the edge of the airport. The house is minutes from the Imperial Highway site where North American's Inglewood factory was once located. Today the area is a vast airport sprawl of hangars and warehouses, gas stations and fast-food outlets, with jetliners approaching LAX in twos and threes, coming in low over the parking lots.

Father and son never spoke of the accident again. Edgar immersed himself in the crowded agenda before him, burying his grief in work. North American was busy with contracts. Before long, Edgar was pushing out new production types and involved each day in wing camber (the exact teardrop form of the wing), wing chord (the wing diameter), dynamic balance, cockpit, cooling and tail assemblies for a new generation of NAA aircraft. He was put to work on two new bombers: the twin-engine NA-21 and the NA-62, soon to become the famed B-25 "Mitchell" medium bomber in which Jimmy Doolittle led the first raid on Tokyo. Edgar had reached the summit of his profession. He was not only an aircraft designer—he was designing planes for one of the most powerful militaries in the world. He was at the Valhalla of his craft.

The design shop on Aviation Boulevard was soon plunged into a new epoch, however, for the Germany Schmued had left behind in 1925 had spawned a monster.

In 1933, Adolf Hitler became chancellor of Germany. Before the decade was out, 70 million Germans had followed him into the fever of Nazism. Nazi Germany, rearming secretly after the Versailles Treaty, grew like a boil and would finally burst in the toxic eruption of the Second World War.

In March 1936, as Schmued began work in California, Hitler marched into the Rhineland and "remilitarized" it. By 1938 he would annex Austria in the Anschluss and a year later he would invade and occupy Czechoslovakia. Finally, in September 1939, advancing Germany's brutal subjugation of Europe, Hitler would invade Poland without pity, fighting horse-mounted cavalry with tanks and touching off another world war. Within a year his army would thunder through northern Europe and arrive at England's doorstep.

The preliminary design group at North American, with Edgar Schmued at the helm, had begun as a small office with four desks and a drafting table. Soon it would expand to meet the accelerating demands of war—and Schmued's reputation would grow with it.

CHAPTER 3

100 Days

It was as humble as a moment of history gets. One day in the spring of 1940, Dutch Kindelberger walked into Edgar Schmued's office, leaned across his designer's desk and asked offhandedly, in his gravelly voice, "Ed, do we want to build P-40s here?" He was referring to the Curtiss P-40 Warhawk, the most popular American fighter in production at the time. His question sounded more casual than it was.

Hitler's invasion of Poland the previous fall had involved some 1.5 million men, 14 armored Panzer divisions and, notably, 1,000 airplanes; Warsaw had fallen within weeks. This new "Blitzkrieg" warfare—a one-two punch of airpower and armor—had left civilian and military leaders around the world looking on stunned.

France and Britain knew they could not make enough weapons to meet the German attack, should it arrive on their own soil. As the storm clouds began to gather back in 1938, the two nations had reached out to Washington for arms. This outstretched hand of the

Old World was the Anglo-French Purchasing Board, headed by a slender, statesmanlike civilian with an intense gaze named Sir Henry Self.

By 1940, Sir Henry was looking for fighter planes. He sought 400 P-40s, but the production line at Curtiss-Wright was already saturated. At the suggestion of the Army Air Forces, he turned to North American. With the Nazi war machine breathing down his neck, Self would need his airplanes quickly—and he knew Kindelberger could produce them. Sir Henry was already familiar with North American's cutting-edge production methods.

The catch was that the P-40 was Curtiss-Wright's creation, not North American's. Kindelberger had no intention of building someone else's airplane; Schmued still harbored dreams of designing his own fighter.

"Well, Dutch," Edgar replied that day to Kindelberger's question, "don't let us build an obsolete airplane. Let's build a new one. We can design a better one," he continued, "and build a better one."

It was the spark and ignition of the P-51 Mustang, the plane that would eventually climb to the top of the battle in the skies. The project launched on that day would become an epochal advance central to all of World War II.

Schmued's response was precisely what Kindelberger wanted to hear, and with it, the design of a brand-new breed of fighter began.

"Ed, I'm going to England in about two weeks," he told Schmued, "and I need an inboard profile, three-view drawing, performance estimate, weight estimate, specifications, and some detail drawings on the gun installation to take along. Then I would like to sell that new model airplane that you developed."

Schmued could turn on a dime. He could have the drawings Kindelberger needed ready at once—for the design was already in his mind. From his earliest years in Landsberg, he had never stopped drawing airplanes. Recently, he had sketched out his ideas for the perfect fighter—what the cockpit would be like, the layout for the engine, the arrangements for the guns. The drawings were sitting in his desk drawer. All he needed was the fuselage, the wings, and the tail. "I made many sketches without knowing what it might end up as," he would say later, "but when the time came, I was ready." He would need just a few days to give Kindelberger what he wanted.

Kindelberger's requirements were simple: "Make it the fastest airplane you can," he told Schmued, "and build it around a man . . . that is 5 feet, 10 inches tall and weighs 140 pounds." The solution was close at hand: Art Chester, a famous racing pilot who had just joined North American as a project engineer, was exactly the right size. With these minimal specifications and nothing but his copy of *Hütte*, a German engineer's handbook, and his little notebook of technical formulae, the Mustang master got to work.

Kindelberger took Schmued's drawings to England, but when he returned two weeks later, he was empty-handed: there was no contract. Undeterred, Schmued began work on a mock-up of the new fighter using paper and plaster of Paris, as the war in Europe unfolded and Hitler's Panzer divisions advanced. "Meanwhile, the British apparently looked at this proposal carefully," Schmued said later, "and concluded that on the strength of North American's past performance with the Harvard Trainer, which was delivered in large quantities [and on which thousands of young British pilots had trained], we could be trusted to come through with a good airplane."

Then, in the early morning of April 9, the Germans invaded Norway, seizing major ports from Oslo to Narvik with land troops and the airports at Oslo and Stavanger with parachute battalions, the first use of paratroops in history. Oslo quickly fell, and Narvik and the necklace of Norwegian port cities followed. That same day, the Germans entered Copenhagen, supported by Luftwaffe squadrons, to occupy Denmark, again under the engine of airpower. Most of Europe was now covered in the folds of the unfurling swastika.

Henry Self knew there was much more to come. Two days after the attack, he dropped his initial insistence on the P-40. He had confidence in Dutch Kindelberger, and it was time to move quickly. Self signed a letter of intent to purchase 400 fighters from North American. With the stroke of a pen, wrote Schmued's biographer, "an Englishman had just launched production of the greatest fighter of the war."

In July a formal contract was signed at last with the Royal Air Force. Three hundred twenty new fighter planes—P-51s—would be delivered by September 1941, with more planes to follow. Each plane would cost $37,590.45. The price was important: Britain was rapidly spending down her cash reserves.

The Battle of Britain raged, and British prime minister Winston Churchill was already pressing President Franklin D. Roosevelt for help. Great Britain was paying cash for arms, but her gold supply was dwindling. In November, FDR would win reelection to an unprecedented third term and move ahead with his "Lend-Lease" program: Britain would soon receive American arms to be paid for not in cash but *in kind* after the war was over. But now, before Lend-Lease, Roosevelt's hands were still tied. He would first unofficially send a shipment of arms to England; then, without congressional approval, 50 surplus destroyers. Beyond that he could not go, and for now Churchill would have to continue spending down his gold for arms.

With the contract signed at long last, Self became adamant about getting his plane fast. Hitler was at his back, with Europe disappearing down a vortex of war. Poland was gone; Norway, Sweden,

Holland and Denmark had been overrun; the Baltics had been occupied by German troops. With Hitler's mechanized armies rushing across the continent, the British wanted their plane designed and built in 100 days.

"Then, as you might say, hell broke loose," Schmued recalled. "We organized a small engineering group and Dutch Kindelberger gave me the right to pick any man in engineering for the project." The dash to bring the plane to life was on. The project was designated NA-73X: the NA stood for "North American," the 73 for the project number, the X for "experimental."

Schmued's team of Mustang men consisted of 14 designers and engineers. There were no women on his team: women engineers in aviation were practically nonexistent. Between 1940 and 1942 the ratio of women to men in aeronautical specialties was about one to 2,000. But as America entered the war, the industry would soon become eager to train and hire women engineers. Curtiss-Wright started a ten-month training program for women called C-W Cadettes. Once employed, though, female aviation engineers were used at a different level from men; they performed only routine tasks—calculations, drafting and illustration. They had no part in the conceptual stages of design.

The Mustang crew expanded at its maximum to a peak of 49 men, then was winnowed down again to 10 or 12 as the aircraft was taken from its first sketches to completion as a working model. Schmued was the leader; he set up working groups for the various phases of the plane's design. There were teams for wings, aerodynamics, instrumentation, fuselage, control systems. The designers worked so closely with the experimental fabrication shop "that the airplane took shape almost as soon as the drawings came out of the blueprint machine."

The design group occupied a spartan room just over Kindelberger's office. One engineer recalled later that "the Mustang materialized out of that smoke-filled room where men knew no hours, where lights never went out, where for days all you could hear was the rattle of paper, the sharpening of pencils, and the noise of men knocking their pipes on waste baskets."

Schmued's team worked every day until midnight—16 hours a day—quitting at 6:00 p.m. only on Saturdays so they had a "weekend." "Sunday was different," one designer recalled, "only because the hum from the adjacent factory was missing."

Schmued's quest to create the world's best fighter plane now became a sprint. A blizzard of some 2,800 drawings would fly off North American's drawing boards by the time the prototype came to life. The NA-73X would have a wingspan of 37 feet and a total wing area of 235 square feet, and every inch of these surfaces had to take into account turbulence, drag, maximum lift, camber (the curvature of the wing) and chord (the centerline of the wing). Every inch would have to be imagined, drafted and finally built by hand. Schmued and his team would translate his earliest concepts into measurements, then drawings, then detailed plans before the prototype would finally take physical shape in the allotted time of 100 days. "It was the greatest team I could ever, ever ask for," Schmued recalled. "People were so cooperative, so hardworking; it was magnificent."

Schmued's concepts for the design of the NA-73X were at his fingertips from all the years of imagineering he had done about aircraft. These were ideas he had thought about for almost three decades. Schmued was convinced his plane must have absolute minimum drag. He would achieve this by using the thinnest skin possible—a thickness of just 0.040 inches. The NA-73X would be the first plane in history whose final form precisely matched its de-sign specifications—not those specifications plus an outer alumi-num coating thick enough to distort the plane's aerodynamics.

To assure that the plane would slip cleanly through the air, the shape of the fuselage (the main body except for wings and tail) must flow in gentle curves. This was the concept of "streamlining," which had been used over the previous ten years, most notably in the Brit-ish Spitfire. Schmued took it one step further: to design exception-ally smooth curves, he would use a series of ever-diminishing cross sections of a cone, which grew narrower in a tapering progression. Again, this idea was new. "The air likes that," he would say. "This is the kind of shape the air likes to touch." It was the first time an airplane was designed with tapering conic sections—second-degree curves. "I laid out the lines myself," Schmued said later, "and it was a first." Schmued had taken aerodynamics past the primitive, rude design of biplanes, then past the streamline of the early monoplane, with the smoothest slipstream conceivable.

The wings would be the greatest departure of all. Already in 1900 the Wright brothers were testing different shapes for an airfoil—or wing—for their historic flight in 1903. Now, through the 1920s, many different shapes would emerge as designers struggled to understand the aerodynamic properties of the airfoil. By 1940, as Schmued sat down to draw the wings of his new fighter, Eastman Jacobs, a young engineer at the National Advisory Council on Aero-nautics, known as NACA, in Langley, Virginia, had defined the gold standard for the airfoil. His wings were already being used every-where: in the British Spitfire and the German Messerschmitt Me 109 and the Focke-Wulf Fw 190. In 1940, two NACA engineers visited North American with talk of their newest airfoil, so revolu-tionary it was then classified as secret and held under lock and key. It was called the "laminar flow" wing; Schmued decided to use it on the NA-73X. It was different from Eastman's other wings and had never been used before.

Eastman had found a way to dramatically improve lift—with a wing that followed exactly the aerodynamics of a teardrop. A wing is shaped with a bulge on its upper surface, so air moving under the wing passes more quickly and air moving over the top moves more slowly. This airflow, which opposes the weight of the airplane, keeps it aloft. The bulge of Schmued's teardrop-shaped "laminar flow" wing was moved farther back from the leading edge of the wing, instead of a quarter of the way back as on older wing designs, significantly reducing drag and turbulence and creating greater lift.

Schmued's work had been foreshadowed more than a century before. Exploring shapes for wings long ago, Sir George Cayley in his trout experiment had predicted the shape of the laminar flow airfoil. One of Schmued's team realized that if you put the measurements of Cayley's trout cross sections over the measurements of the laminar flow wing, they were identical. "So God invented the laminar flow airfoil way before us," he remarked. "This helped us to be humble."

Drag and lift: these were paramount in every aspect of the new fighter plane. Minimum drag, from the carburetor intakes in the nose to the wingtips. Punctilious care to the 1/50,000th of an inch for many of the parts of the prototype. Schmued and his men pressed on into the wilderness of aeronautics, bringing the NA-73X slowly into existence.

Laminar flow wing. Second-degree curves, a Meredith radiator, screws flush with the metal skin: all went into the novel genus of fighter plane.

The team raced on, working at fever pitch, rushing down to their approaching deadline. There were problems with the landing gear. Too narrow. Solved. Problems with balance between the tail and the front of the plane. Solved, another first. This was called dynamic balance; today it is a principle of all aircraft design. Flap problems, glitches with the elevators, gunnery issues: all were worked out by Schmued's crew. And so the body of a new plane emerged from the drawing boards of Schmued's team like a Frankenstein, or the mythical Jewish superman, the golem, who rises from mud and twigs, coming to life.

Schmued oversaw the constellation of details and problems with his trademark calm. His staff found him immensely likable. With his stocky frame and dark looks, his heavy German accent and old-fashioned European manners, he was steady and supportive. Sometime each evening, one of his team remembered, "[he] would drop around to check our progress. Ed never pressured us . . ." Encouraging the men, he was always a good listener.

With the design of the plane finished, magnesium castings of the aircraft's parts must be made from the drawings. One part was so huge, only one foundry would touch it. "When they saw the drawing, their boss nearly collapsed," Schmued recalled. "I sent one of our designers to stay . . . as long as necessary until the casting was made." It was "the biggest magnesium casting anybody could think of. . . . The man that I had sent there took the hot casting and put it in the rumble seat of his car. When he delivered the casting, he noticed it had burned a hole in the rumble seat's floor."

As the summer wore on, his workers grew exhausted and mistakes crept into the work. Schmued rarely showed impatience. On one occasion, a team member used an incorrect calculation to render a piece of the plane's landing gear. The man feared the entire project would be delayed because of his error. "I told him let's not worry who makes a mistake, let's see what we can do to fix it," Schmued recalled. "When you rush a design through like this, you are inviting mistakes. And they do happen. In every case, it is important not to worry who did it, but how to fix it."

But even the boss had his testy moments in the sprint to complete the new NA-73X. Vern Tauscher, an engineer on the project, remembered an incendiary episode near the completion of the Mustang prototype. Not long after the first plane was finally assembled, the test pilot reported that the brakes on the landing gear did not have enough pressure; Schmued told Tauscher to boost it. Tauscher did so. The test pilot tried the brakes again—the pilot said they still needed more pressure. Tauscher boosted pressure again; the pilot said it still wasn't high enough. The tension among the men on the runway began to rise. Schmued stewed, simmered in silence, then finally erupted. In a voice that carried across the airfield, he bellowed at Tauscher, "I should kick your ass all over this airport. I should fire you. Now you go over and give me more pressure."

Tauscher kept his cool. He walked over to the NA-73X and set the pressure once more without argument. The pilot then taxied away, took the prototype up to takeoff speed—and the wheels burst into flame. The pressure had been too high. A fire truck was standing by to put out the fire, and the test continued. No one bothered Tauscher about the brake pressure after that. At his next wage review, the boss raised his salary.

By the time the prototype for the NA-73X was finished, 41,880 engineering man-hours had been expended on the project. The team had worked fast and furiously through July and August, and in September the airframe was ready. They had missed the British deadline by just two days. "When 102 days were over everybody collapsed," Schmued recalled. "We took a few days off because everybody deserved it." Unfortunately, its engine, made by Allison in Indiana, was not ready. Schmued remembered, "The Allison people . . . told us that since nobody ever designed a fighter in 100 days, why would you need an engine in 100 days?" But North American had done the impossible. Schmued parked one of his engineers at the front door of the Allison plant in Indianapolis with the order: "Don't come home without an engine." Delivery of the engine would take another 18 days.

Finally the prototype was complete.

In gleaming aluminum and magnesium, it looked like a hawk, with the squarish head of its cockpit descending to a sharp beak of a nose, full, balanced wings and a tapered, narrowing body; it seemed almost primitive, with fuselage forms and outlines that had never been seen before. It combined speed and maneuverability in its 32-foot-3-inch length.

With its launch, Edgar Schmued would join the star field of aviation design immortals: Kelly Johnson of Lockheed, who had shucked out the P-38 Lightning; the British designer R. J. Mitchell of Supermarine Aviation, creator of the Spitfire; Alexander Kartveli of Republic Aircraft, master of the P-47 Thunderbolt; Jack Northrop of Northrop Corp., father of the P-61 Black Widow.

October 26, 1940, was D-Day for Schmued's prototype.

That morning, 97 North American employees—engineers, plant workers and Schmued—gathered in a close cluster at Mines Field in the California autumn haze to witness the first official test flight of the NA-73X. The aluminum-alloy prototype, unpainted except for the black license number NX19998 on its wings, sat on the runway, its sculpted contours and sparkling Duralumin features catching the morning sun. After some last-minute consultation, the test pilot Vance Breese got into the plane and fired up the 1,120-horsepower Allison V-1710-39 engine. It had rained the previous night, and the NA-73X spun up mist as it taxied down the runway and took off. Breese roared to 3,000 feet, swung and veered, climbed and raced through the California altitudes. He sped west toward the coast, peeled off south and let go the throttle. The plane that would fly straight to the center of World War II was flying for the first time. Gazing skyward, the crowd on the ground let out a cheer.

It was a short flight, a truncated maiden voyage. In the next days, Breese flew the prototype again and again, checking and re-checking: rate of roll, elevator action, aileron action, fuel system, windshield pressure, stability and a dozen other measures. The NA-73X aced all its trials, exceeding Schmued's expectations. Hardly any changes would need to be made. A new specimen of aluminum raptor was aloft, faster and more nimble than any fighter ever built.

CHAPTER 4

Death from Above

In the summer of 1940, as Schmued and his team dashed toward their deadline, ominous events had engulfed Europe. France fell in six fast weeks. The French Army, once thought to be the finest fighting force in existence, collapsed in an astounding 42 days in May and June like a rickety scaffolding toppling down, arguably the greatest military disaster in history. Airpower had driven its defeat.

In May, the Germans had raced across their western border at Aachen and Liège into the Low Countries, launching Fall Gelb (Operation Yellow), the offensive against Western Europe. Within days they had tramped across Belgium, Holland and Luxembourg with the twin iron rams of their "Blitzkrieg"—armored Panzer divisions and airplanes. German mechanized units forged into the dense Ardennes forests, covered by a force of 1,000 Luftwaffe planes; now they breached French defenses along the Belgian border. The legendary Nazi tank ace, General Heinz Guderian, broke through the Ardennes with his XIX Panzer Corps; in one hurtling charge he raced to the

English Channel coast at Abbeville, pulling an armored noose around the entire northeast corner of France from the western Belgian border to the sea. Trapped inside this choke hold of Guderian's mechanized Panzer units were 500,000 Allied troops at the port of Dunkirk—the fighting best of the armies of Europe. In the next few days, the Royal Navy and a heroic ragtag civilian fleet of 700 private English fishing and pleasure craft managed to evacuate most of the ensnared troops. The Battle of France was effectively over.

In June, the Nazis launched their final offensive to take Paris and destroy what remained of the French army in the field. Tides of Luftwaffe squadrons joined armored units and infantry to surge through the French midlands, rolling west and south. On June 14, Paris fell. Nazi storm troopers goose-stepped down the fabled Avenue des Champs-Élysées as an immense swastika was unfurled from the Arc de Triomphe. By the twenty-second, an armistice was signed and France had surrendered. The republic of King Louis, Joan of Arc and Napoleon was gone.

From the green Suwałki expanses of Poland, to the snowcapped peaks of Norway, to the lush rows of France's Burgundy vineyards, death from above had swept across Europe. Germany had ushered in a new age of air war—bombing, strafing, dogfighting, attacking the ground from the air. In Poland, a vast fleet of Luftwaffe planes had backed up the Panzer and infantry divisions invading Warsaw. In Holland, Belgium, Luxembourg, Denmark, air assault had led the attacks. Rolling through France and Scandinavia, Hitler's Blitzkrieg had teamed murderous armored Panzer divisions with the punch of airpower: armor, air; armor, air. The nature of warfare had been changed by Hitler's Luftwaffe, the German planes devastating everything in their path. Manned, powered flight, barely one generation old, had been key to the lightning advance of the Nazis. Their dark raptors, the Messerschmitt Me 109, the Ju 87 "Stuka" and the Dornier Do 17, were like heat-seeking missiles, the rockets of their day. They were the new lords of war.

By July, as North American commenced work on the P-51, Hitler had locked up all the continental expanse of Europe. Now, from Paris to Amsterdam and the boulevards of Warsaw, the Old World and all the mainland lay under the jackboots of the Nazis. Hitler then swiveled his guns, turning them on the last people still unbowed by his forces: the British, in their isolated island home. The German attack, again, came from the sky.

The tense, taut struggle that July in the clouds over Essex and Kent played out in the fast, desperate duel called the Battle of Britain, in which 2,936 pilots of the Royal Air Force, outnumbered by Germans four to one, beat back the Luftwaffe in 113 frenzied days and saved their kingdom from the incubus of Germany's Führer. The heroic RAF airmen were led by the quiet, stubborn and brilliant tactician Air Chief Marshal Sir Hugh "Stuffy" Dowding.

The Battle of Britain was fought at laser speed and in almost gymnastic combat. Between July and October, RAF Spitfires and Hawker Hurricanes knocked the invading Luftwaffe planes out of the sky—serpentine Heinkel He 111s, froglike Dornier Do 17s and ash gray Messerschmitt Me 109s. The airmen's tactic was to climb high over the attacking Germans, get the sun behind them and then swoop down out of the glare to pounce on the foe. Then the dogfight would begin, Hurricanes and Spits diving, banking, plunging, climbing in among the waves of German fighters and bombers.

With a top speed of 374 miles per hour, the Spitfire is remembered as the star of the Battle of Britain, but the Hawker Hurricane was the real quarter horse of the engagement, bagging 70 percent of the kills. These were the chargers that raced through the battle smoke of England's skies in the heat of that crucial summer; they fought in the knowledge that every German assailant who got through was gunning for their people, their villages, their homes and their families. In July, the Germans sent more than 2,500 planes against England; the British defended their island with just 640 fighters.

The British were better pilots, but Dowding also had a piece of mechanical sorcery up his sleeve: radar. The Germans had nothing like it. The brainchild of Sir Robert Watson-Watt, the great network of stations ringing the chalk cliffs of the Channel and the southeast coast was called "Chain Home"—a series of towers, like a wall of flat panel antennae, that could track enemy planes. The shroud of the radar waves could reach 100 miles out and pick up German attackers as soon as they began forming up over their bases in occupied France.

Linked by secure telephone lines to the central command system, Chain Home radar stations kept a constant watch over the French coastline. When a formation was sighted, each station, manned by a WAAF of the Women's Auxiliary Air Force, would transmit the apparent range and direction, converted into map coordinates, to a "filter" room in Fighter Command Headquarters at Bentley Priory, north of London. Here the information would be received by the underground plotting room, where markers were moved around a board with a huge map displaying enemy formations. Officers watching the enemy markers would guess at the bombers' targets and alert the Fighter Command groups. From there the information on expected Nazi raids went to designated "sector operations rooms," which would order individual fighter squadrons to scramble. This was Air Chief Marshal Dowding's great tin umbrella, spoking out across England's skies to allow the RAF to "see" the plan of battle and direct it.

It was a close-run thing. On September 7 came the ultimate battle; that morning 2,000 German planes, more than 800 of them bombers, droned in over England. The RAF pilots fought them for nine full hours with a loss of 34 planes to a tally of 75 Germans destroyed. Watching the fearsome scrimmage with Air Vice Marshal Keith Park in the operations room of RAF 11 Group in Uxbridge that day was Prime Minister Churchill, waiting tensely for the outcome. As wave after wave of Nazi bombers and fighters approached the coast to drop their loads on English turf, the prime minister asked anxiously what reserves remained. Quietly the air vice marshal replied, "There are none." He had put up his last plane.

At the end of the day, the exhausted Churchill ducked into his car to be driven home with General Hastings Ismay, one of his closest advisors. He brooded silently, lost to his thoughts. "Don't speak to me," he said to Ismay. "I've never been so moved." And then Ismay heard Churchill mumbling to himself: "Never in the field of human conflict was so much . . ." He was forming his historic testament to the RAF airmen, which he would repeat a few days later before Parliament: "Never in the field of human conflict was so much owed by so many to so few."

The British pilots were the victors. Over three months and three weeks they shot out of the sky a total of 1,887 German planes while losing only 1,023 of their own aircraft. Five hundred forty-four RAF airmen lost their lives while a swarm of 2,600 German aircrew went to their deaths. Poles, Canadians, Irishmen, New Zealanders, Czechs, Belgians, Americans, Frenchmen and others, too, flew for England through that "bright and terrible" summer and fall.

In October, as the completed NA-73X took wing for the first time in California, Hitler gave up on his goal of destroying the RAF and on Operation Sealion, his plan to invade Britain. The war of attrition had grown too costly: neither he nor his Luftwaffe chieftain, Reichsmarschall Hermann Göring, had the stomach to risk the wholesale loss of their air force. So they turned to another gambit—crushing the will of the British people by bombing their cities, wearing them down to despair and perhaps surrender. The Nazis returned with their fast bombers, medium bombers and fighters droning overhead in the hundreds. Once again airpower was the weapon of choice.

London was raided for 76 consecutive nights. The raids were staged at night to maximize shock, incendiary bombs lighting the way for wave upon wave of Luftwaffe Heinkels and Messerschmitts. In central London, citizens took to the underground subway stations and government "Anderson" shelters for refuge from the assaults, arranging bedding and personal effects for the tense, long nights. Soon the bombers hit the docks and mazelike wharves of the East End, where the poorest lived, and night after night the waterfront was targeted. The stolid East Enders stood up to the onslaught with defiance and grit and became famous for what was termed "war-time spirit." Buckingham Palace was bombed, Westminster Abbey and Parliament were damaged, but Londoners hung up signs in their shops declaring "Business As Usual."

The city of Coventry was attacked by 509 bombers in a raid so severe, it generated the term "Coventrize"—to stage an aerial attack so devastating it creates mass havoc. Some 1,400 high-explosive bombs and 10,000 incendiaries were dropped on the city in a bombardment that touched off a firestorm. The magisterial turrets and spires of Coventry Cathedral, which had stood since the fourteenth century, were reduced to rubble, so that only the soaring west tower remained.

"The Blitz," as the aerial campaign of Hitler's bid to crush the British would be known, was relentless. For 267 days, from September 1940 to May 1941, the British suffered a continual barrage of London and other strategic cities across their kingdom. London seemed to many like a desolate tundra or wasteland, almost devoid of relief. Bomb shelters were overcrowded and had poor sanitary facilities. Danger was ever present from unexploded bombs. "For Londoners, there are no longer such things as good nights; there are only bad nights, worse nights, and better nights," Mollie Panter-Downes wrote in her memoir of that winter. "Hardly anyone has slept at all in the past week."

The Blitz crept through three seasons. The RAF learned to fight at night; the repair crews dug out by day; the British people hung on. Forty-three thousand were killed in the campaign, including 5,028 children, and still the British did not lose their resolve. For the first time, Hitler had faced the only solid opposition the Nazis had encountered in all of Europe. This was not Poland, not Norway, Belgium, nor even France. The British did not submit.

Finally Hitler gave up his hope of driving them to despair. He pulled back his Luftwaffe: its losses had become catastrophic, from 28 aircraft a month in January 1941 to a staggering 124 per month by May. After months of deliberation, he turned his sights to the east, unleashing Operation Barbarossa, the invasion of Russia. In June 1941 he sent three million men, 150 army divisions supported by 2,500 aircraft, out into the immense expanse to the east, where they would soon be lost in the icy grip of the vast Russian winter.

The British pilots who had won the summer clash had once more prevailed. They had defeated the Germans in both the Battle of Britain and the Blitz and, for the moment at least, driven them from their shores. And as she had almost been doomed by airpower, now Britain had been rescued by it. She was saved not by her vaunted Royal Navy, supreme for more than three centuries, nor by her army, the builder of her empire from Khartoum to Bengal, but by her pilots and her air force, then only thirteen years old.

CHAPTER 5

A Turbine like a Typhoon

The harrowing battle for Britain's survival, encompassing ten months of strain and struggle, was finally won by young pilots who outdid their numerically superior Nazi adversaries, superlative aircraft like the Supermarine Spitfire I and the Hurricane Mark II, and the flint and foresight of Air Chief Marshal Dowding. But the real hero of Britain's twilight stand was neither a pilot nor a general. The real hero of the engagement was a 1,640-pound liquid-cooled V-12 piston engine, the Rolls-Royce Merlin 61, the raging bull of so many British warplanes.

Produced by the legendary English coach maker Rolls-Royce, and named, like its other warplane engines, for a bird of prey, the Merlin was a supercharged powerhouse that turned like a Tesla dynamo and had the power of a typhoon. In the end it would go into some 40 aircraft flown by the RAF throughout the war and would also transform the P-51, projecting it into the battle the Mustang would lead to victory. Without this thundering turbine of screaming valves, cams, rocker arms, push rods, bearings, tappets and springs, the British pilots could not have saved their island, and the P-51 would not have entered the decisive clash of World War II.

Though its name conjured up a mythical wizard, the Merlin was in fact named for a type of falcon so fast it could outfly any attacker. It was invented and developed by a Laurel and Hardy odd couple, two men of such improbably different beginnings, their story was almost a Pygmalion stage drama in itself.

Henry Royce, one of the greatest tinkerers and inventors in history, grew up dirt-poor, had only one year of formal education and survived his childhood by working as a newsboy, a telegraph messenger and a railroad apprentice. Like Edgar Schmued, he grew up fascinated by motors and machinery.

Charles Rolls was the prince to Royce's pauper. The son of an aristocrat, educated at Eton and Cambridge, Rolls loved speed and was fascinated by racing. He spent his youth flying balloons and early powered aircraft, and regularly entered motor competitions. He eventually began selling luxury cars to the wealthy.

Royce's first venture was in electricity, the phenomenon of the day. At his workshop in Manchester, he made and sold light switches, fuses and doorbells. Soon he went on to manufacture some of the best large-scale industrial electric cranes in the world: one client was the Imperial Japanese Navy. In 1902 his interest turned to the emerging automobile, then the province of French builders—de Dion, Delahaye and Panhard—and in 1904 he produced a fine little car: the Royce Ten, a two-cylinder, ten-horsepower jewel that ran as efficiently as a hand-wound Chelsea clock. It had an entirely new kind of carburetor, a wonderful smooth suspension, three forward speeds and a clutch lined with leather.

Charles Rolls, by then a leisured aristocrat and daredevil trying to make a go of it selling fine cars, mostly Peugeots and Panhards, heard about Royce's little marvel and quickly became obsessed with it. In May 1904 he traveled to Manchester to see the Royce Ten for himself. He was astounded by how smoothly the engine ran; he was captivated by how elegantly the carriage rolled; he was spellbound by the perfection of Royce's little runabout. The two signed papers formalizing their partnership, and in 1906 Rolls-Royce Ltd. was born. The company would go on to build the finest motor coaches in the world, traveling saloons for the right sort of gentleman: the Silver Ghost, the Phantom I, the Silver Wraith, the Twenty, and later the Corniche and the Silver Cloud. In the early 1900s, these cars defined performance and luxury.

Soon the Great War would consume all of Europe. The upheaval created a new market for Rolls-Royce when the British government drafted the company to produce aerial engines. By 1918 they had made more than 60 percent of all British plane engines used during the war. Royce himself designed a new engine for France's air force and built it in six months. This was the Eagle; the Falcon, the Hawk, the Condor and the Griffon would follow.

After the Great War, the demand for aircraft—and aircraft engines—thundered on. This was a prodigious time for aircraft development, a boom era of rapid change in aircraft design. The public was obsessed with flight and part of that was an obsession with speed. Air races were all the rage.

The Schneider Trophy race, which would prove so important to the development of the Merlin engine, was a biannual air race started in 1913 by the French financier and balloonist Jacques Schneider for seaplanes, the fastest aircraft of the day. The race delighted and enthralled crowds for almost two decades and was one of the most prestigious events on the aviation calendar.

Winning it was the dream of Britain, the U.S. and Italy, then the leaders in seaplane design. Britain had won the race in 1927; to defend the trophy in 1929, Supermarine Aviation Works decided to use a new engine, an experimental one from Rolls-Royce. Henry Royce began to design it in 1928, one day sketching his concept in the sand as he walked with his engineers on the beach near his home in West Wittering, Sussex. The "R" engine won the 1929 trophy, flying at 357.7 miles per hour, a new world record.

Then fate nearly carved Britain out of the running. In the grip of the Great Depression, the government of Ramsay MacDonald pulled funding for an English contender in 1931. Enter the British press—which headlined the unpopular decision—and an eccentric but patriotic zealot named Lady Lucy Houston. Working-class daughter of a box maker, at 16 a chorus girl with "impudent speech and a tiny waist," a fresh-air fiend and nudist who had married into aristocracy three times, "Poppy" Houston stepped forward and announced she would put up £100,000, $8 million in today's money, for a British entrant. The seaplane was built, with an even more advanced "R" engine, and won the 1931 race, setting a new record of 407.5 miles per hour and keeping the victory cup in perpetuity for England.

Beyond the groundbreaking Rolls-Royce engines, the seaplanes themselves were also critical to British victory in the Schneider race meetings. Britain's streaking seaplanes—the S4, S5, S6 and S6B, designed by a young and already storied prodigy in aviation named Reginald J. Mitchell—were the bearers of England's glory. By his twenties, Mitchell was the chief designer at Supermarine Aviation; his sleek monoplanes in aluminum, not wood, set new speed records. He began to design warplanes for the RAF, developing the aircraft that would soon take flight in 1935 as the Spitfire with the Merlin engine on board.

The Schneider Trophy race was key to the development of the Spitfire and its Merlin engine, but it would be vital, too, to the development of all the Merlin engines that followed: the Merlin Mark I, II, III, the Merlin XX, the Merlin 45—and the Merlin 61. Henry Royce's work on the "R" would lead to the powerhouses of the Battle of Britain and then to the unleashing of the P-51B Mustang on the demons of the Luftwaffe. At the heart of these advances in power were the supercharger, the turbocharger and the injection of pressurized air.

An airplane engine is exactly the same as a car engine—just like a Dodge, Jeep, Nissan, Ferrari, or Porsche—only more powerful. The internal combustion engine uses a fuel-air mixture and electricity from a spark plug to create a controlled explosion in each engine cylinder—6, 8, 12—the force of which drives pistons back and forth, turning a crankshaft at the base of the engine that spins wheels, or propellers. Thus, gasoline explosion pumps piston, which turns crankshaft, which turns wheel or propeller in a tame union of force and motile drive. Except that the Merlin 61 could hardly be described as tame. It was a thundering Miura bull.

A typical ship engine runs at 70 revolutions per minute. A medium-range diesel runs at 1,800 rpm. The Merlin engine screamed and shrieked at 3,000 rpm. A standard car engine develops about 300 horsepower. A typical 18-wheel semitruck rig develops about 500 horsepower. The Merlin 61 delivered 1,490 horsepower. It ran like Craig Breedlove setting the land speed record.

Aerial engines, however, are highly susceptible to power loss in the thin, ethereal atmosphere at tropospheric altitude because of loss of pressure to the engine's feeder air. Superchargers are used to boost the pressure on this "charge air" to recover performance. A turbocharger, identical in almost every action, does exactly the same thing—but uses exhaust gases instead of clean air to boost pressure. To chill the superheated feeder air in the lines and conduits of the system, an "intercooler" freshens the charge. Equipped with such a system, an airplane will not lose power at altitude. With the Merlin 61, which had two superchargers in a two-stage setup, it was possible to fly at 437 miles per hour at an altitude of 25,000 feet.

Thus the heart of Rolls-Royce's successful Merlin engines was the supercharger. But the early superchargers did not produce enough power, a fatal flaw in combat. In the late thirties, as war loomed and the RAF was seeking more powerful engines to meet the German threat, a young aeronautical engineer named Stanley Hooker would solve that problem. In 1938, Hooker, a newly minted Oxford PhD with a dissertation on wind tunnels, arrived at Rolls-Royce without a specific role. After several days reading the newspaper in his small office and wandering the halls, he happened on a colleague studying sheaves of paper with figures on the supercharger. Taking the calculations back to his office, he emerged a couple of days later with recommendations for improving the supercharger so it could produce more power.

Hooker's calculations were seminal. His conclusions were so brilliant that he was immediately put in charge of supercharger development at Rolls-Royce. The result was increasing power at higher and higher altitudes from then on through the many incarnations of the Merlin—which went from 1,000 horsepower as the war began to over 2,000 by its end.

The wizardry of this engine, designed near Robin Hood's Sherwood Forest in Nottinghamshire, would not only deliver England in its hour in 1940, but would also make possible the eleventh-hour triumph of the P-51 Mustang when it thundered into battle in 1944.

All of this was unseeable in Nottingham in England, and in California in America, so early in the growing aerial clash that would soon rule the world at war.

CHAPTER 6

The Butcher Bird

In the summer of 1941, an RAF test pilot flew to California to try out the new P-51, just off the production line at North American. He was well qualified for the job: Michael Crossley was a wing commander in the Royal Air Force and a hero of the Battle of Britain who had shot down nine German planes. There was one awkward problem: Crossley was six feet, two inches tall; the aircraft had been designed for a man five feet ten. With his knees pulled up under his chin, he took off and threw the plane through an hour of rolls and dives. The Mustang performed perfectly. It was ready to go to war.

Manufacturing was well underway. Even before the brilliant flight test of the NA-73X the previous fall, North American had been pushing forward with production engineering and tooling to mass-produce the plane. The prototype had been brought to life from drawings by skilled craftsmen, but now each part of the plane had to be broken down into simple operations to be moved on an assembly line by unskilled workers. One hundred fifty-four thousand engineering man-hours and six months later, the new P-51As were rolling out of the Inglewood factory and into the California sun.

The new Mustangs were flown to the East Coast by civilian ferry pilots. There they were either shuttle-flown across to England or disassembled, crated and loaded onto freighters to travel in cargo convoys on the perilous North Atlantic route, the hunting grounds for Hitler's low black U-boat submarines. Wolf packs of U-boats hit the Allied convoys like sharks in a feeding frenzy, in battles lasting for as many as five days. In October of 1941 the Battle of the Atlantic—with two enemies, the Type VII U-boat and the moaning, howling cold—was going badly: that month the Nazi subs sank 32 freighters totaling 154,593 tons. At least 20 Mustangs were lost early on from these heaving convoys crossing the blank 3,500 miles of the frigid North Atlantic.

The first Mustang, bearing British serial number AG346, arrived in Liverpool on the River Mersey docks on October 24, 1941; it was reassembled at Speke airfield nearby and flown for the British press to great fanfare. Its appearance signaled something new. The North American Aviation plane had a strangely futuristic look, almost dronelike, different from the tubular, tapering length of other fighters of the day. Most fighters had a narrow, elliptical fuselage, slender at the rear; the Mustang had rounded, practically oval lines, almost like a discus that would slip through the air without kicking up any turbulence. The nose of most fighters looked like a cigar stub; the P-51's narrowed to a thin point. The cockpit was long and sculpted, like a glass bubble, blending into the funnel of the slender fuselage so there was only one master silhouette to the plane. Just below the belly between the wings was the huge air intake, giving it a dangerous, underslung look. The P-51 looked as if it had rocketry in its soul.

The first RAF unit to receive Mustangs was 26 Squadron; other RAF outfits soon followed, including 241 Squadron, then 2, 16, 268 and 613 Squadrons, and two Royal Canadian Air Force Units, 414 and 430 Squadrons. British pilots raved about it: faster than a Spitfire up to 25,000 feet, it could also outdive the Spit. And because North American had designed it with plenty of fuel, it had more than three times the Spitfire's range—up to four hours, more distance and endurance than any other fighter. It was the finest and most versatile American fighter aircraft yet to reach England. But there was a problem.

The new P-51As had a serious drawback, a compromising flaw, and it was not Edgar Schmued's design. The problem was the engine. On testing at the proving grounds of the RAF Air Fighter Development Unit (AFDU) in Duxford, near Cambridge, and elsewhere, when the Mustang was flown and compared to British and captured German fighters, the American Allison engine performed well at low altitudes, but its power fell off dramatically the higher the fighter was flown. At altitudes above 25,000 feet, the plane became difficult to handle. Not only was it slower; it was also harder to maneuver. As it was, equipped with its original Allison engine, the plane would never be a high-altitude fighter.

At first the British didn't find this important. By then the Battle of Britain had been won, and there was little use for planes that could fly high. The RAF needed a plane capable of both ground attack—strafing German fortifications and installations, railroads, highways and canals—and reconnaissance. Fast and maneuverable at low altitude, the P-51A was ideal for both roles. Fitted with a K-24 camera behind the pilot that pointed backward and down to the left, the aircraft could take clear pictures up to 9,000 feet. It collected hundreds of invaluable photographs of German positions in France.

But that summer, the face and fortunes of the RAF's war were changed. In August 1941, a new shape was spotted by RAF airmen over the French coast: a blunt-nosed airplane with black Nazi crosses on its fuselage and tail. At first RAF Fighter Command thought it might be an American P-36 captured from the French. But soon they realized the Germans had something new—and deadly. Sightings of this fighter, with its new outline of squared wings and a more tapered, reptilian fuselage, began to increase over northern France. Back at their fighter bases, RAF pilots were asked to draw sketches and side views of the menacing new plane for British intelligence.

Battle exposure only confirmed what RAF commanders already feared. Soon the new German plane was inflicting heavy casualties on the Spitfire, with few losses in return. Faster in a climb than an Me 109 and far more stable in a vertical dive, it was faster and more agile than the Spitfire, with excellent visibility and heavy armament. It was a formidable opponent in a dogfight. As the great Spitfire ace Douglas Bader, who flew in the Battle of Britain without legs, would recall: "The plane certainly gave the British a shock. . . . It out-climbed and out-dived the Spitfire. Now for the first time the Germans were out-flying our pilots."

The Nazis were indeed wielding a new weapon. What they had was the Fw 190. Conceived in 1939 by Focke-Wulf's chief designer, Kurt Tank, the fast, nimble 190 was powered by a BMW radial engine developing 2,100 horsepower (compared to 1,440 for the Spitfire and just 1,185 for the Hawker Hurricane) for a maximum speed of 406 miles per hour. Mounting two machine guns and up to four 20 mm cannons, it was an exquisite instrument of death. The German pilots called it *der Würger*—the Butcher Bird.

Months after the Fw 190 first appeared, a German pilot's error dropped a treasure into British hands. The pilot, Oberleutnant

Armin Faber, disoriented after a dogfight with Spitfires over the Channel, spotted what he thought was a Luftwaffe airfield on the Cotentin Peninsula in Normandy. Instead, he landed at an RAF air base in Pembrey, Wales. Taxiing to a stop, Faber was stunned when an RAF mechanic jumped on his wing and pointed a pistol at his head. His captured Fw 190 was a windfall for the British. The airframe and engine were quickly dismantled and analyzed. Then the plane was reassembled and flown to the Air Fighter Development Unit in Duxford near Cambridge. Testing there confirmed the RAF's worst nightmare: the 190 was superior to anything they could put up. It had greater firepower and greater maneuverability than its predecessor, the Messerschmitt Me 109; it quickly proved superior to the Spitfire Mark V and would maintain superiority over all Allied fighters for another critical year. Behind thick barricades of sandbags at the Air Ministry in London, the staff began to dread each passing month. They knew they must now find a new fighter to match the Luftwaffe's predatory plane.

No one thought of the Mustang. It languished in the shadows, unnoticed in its supporting role of reconnaissance, strafing and ground attack. And there it would have stayed until the end of the war but for the merest lucky roll of dice.

One day in April 1942, a key RAF officer picked up the phone. Wing Commander Ian Campbell-Orde, head of the AFDU, had just flown the P-51A for the first time. He was intrigued by its maneuverability and speed. On a hunch, Campbell-Orde placed a call to a test pilot he knew at Rolls-Royce, the engine maker for so many RAF aircraft. He asked his colleague to come up right away and fly the new plane. The test pilot was Ronnie Harker.

Harker came the next day. He made a lark of it, hiring a car from a local garage and taking his wife with him, along with a fuel engineer from the Rolls-Royce proving grounds at Hucknall. They arrived in Duxford on April 30, a bright spring Thursday. That day, Harker climbed into the cockpit of the Mustang, fired up the Allison V-1710-39 plant and took the P-51A aloft. He disappeared into the clear, void dome over Cambridgeshire, driving up into the clouds. He ran through a series of turns, dives and climbs, flew at full speed and threw the plane around in the skies. The plane flew handsomely. It handled marvelously. Its speed, he noted, was 35 miles per hour faster than a Spitfire's at the same power. He rolled, swung, banked, chased south to Heydon. Hurling the Mustang across the vacant blue firmament, Harker had an epiphany. If the thrust of a powerful Merlin engine could be substituted for the Mustang's weaker Allison, the plane would outstrip the Spitfire. With the Rolls-Royce Merlin 61 installed, he thought, the P-51 would outperform even the Focke-Wulf 190. And with its prodigious internal fuel capacity, several times the Spitfire's, the fighter would be able to escort bombers flying deep into Germany. In short, the North American Mustang would be a matchless ace trumping anything the Germans could put up. Harker believed this plane could be the weapon the British so badly needed. He was convinced they had their javelin.

CHAPTER 7

High Noon

Harker landed the P-51 on the tarmac at Duxford and climbed out of the cockpit. Thrilled with his flight, he wasted no time. He went directly to find Wing Commander Campbell-Orde. He told the officer in no uncertain terms what a superb plane the Mustang was and the astounding prospects it offered if fitted with a more powerful Rolls-Royce engine. Campbell-Orde's response was immediate: he agreed on the spot with Harker's assessment. The next day, Harker sat down and wrote a prophetic letter to his superiors. He recommended replacing the Allison V-1710-39 with the more powerful Merlin 61. "The point which strikes me," he wrote, "is that with a powerful and good engine like the Merlin 61, [the Mustang's] performance should be outstanding, as it is 35 m.p.h. faster than a Spitfire V at the same power."

Harker's next step was audacious. "Full of excitement," he would write in his memoir, Harker went straight to R. N. Dorey, the general manager at the Hucknall proving grounds, and at once enlisted him with his ecstatic test reports on the North American Aviation plane. Dorey saw immediately the dramatic opportunity presented by Harker's tests; he encouraged him to go directly to the top. Harker next drafted a report to the Rolls-Royce brass urging them to refit a couple of Mustangs with Merlins and test them. From that report, the idea of a Merlin 61 transplant would go viral.

Word now came back that Ernest Hives, the director of Rolls-Royce, later Lord Hives of Duffield, would see him. "Churchillian in physique, mentality and powers of leadership," Hives was the godfather of Rolls-Royce, a man who had climbed the ladder from shopworker to chief engineer. He did not suffer fools. Harker prepared for the meeting as if for a final exam. He contacted Witold Challier, a brilliant Polish mathematician who had fled his homeland in 1939 and was working at Rolls-Royce as a performance engineer. He asked Challier to produce a set of "performance curves," graphs that would estimate the top speed and rate of climb for the Mustang with a Merlin engine compared to a Spitfire.

The results of Challier's calculations were staggering. Slide rule in hand, he produced a set of graphs forecasting the speed of the Merlin Mustang at all altitudes. Taking into account flight characteristics such as propeller diameter and efficiency, correct gearing figures and intercooler performance, Challier showed that the speed of the new genus would be 345 mph at sea level, rising to 410 mph at 13,800—and a breathtaking 441 mph at 25,600 feet. His calculations would later prove on the mark, if too cautious. The bird would fly like a field arrow, definitely outstripping both the Spitfire and the Fw 190. Challier had confirmed what Harker suspected: a superlative performance would come from mating the Merlin's fury with the Mustang's streamlined airframe.

Armed with these charts, Harker prepared to call on Hives at his office in Derby. An affable man with prominent, beaky features and a lanky frame that gave him a certain resemblance to a large bird, Harker had joined Rolls-Royce straight from boarding school as an apprentice in 1925. He had been besotted with motorcars as a boy, but when his father took him to the Hendon Air Pageant two years later, his dreams changed. He learned to fly, joined the RAF Auxiliary Air Force and became Rolls-Royce's first test pilot.

"With all the enthusiasm and conviction I could muster," he would later recall, "I told [Hives] what a splendid aircraft the Mustang was and how it was needed by the RAF and could we have one at Hucknall to do a conversion?" Hives was won over. Never one to waste time once his mind was made up, he lifted the phone to Air Chief Marshal Sir Wilfrid Freeman, vice chief of the Air Staff, and asked for a few of the Allison P-51s. Freeman did not hesitate. He was the ideal quarterback for the Merlin assignment, involved in the development of virtually every aircraft brought along by the RAF between the wars. From then on, he would use his power and access to move the project along. When Hives wanted three Mustangs for conversion, Freeman would ask for six. When Hives wanted 250 sets of conversion parts, Freeman would find him 500.

The first Allison Mustang arrived at Hucknall for conversion in June; preparations began immediately. In all, five Mustangs would be delivered from the 138 that had arrived in Britain by this time. The actual engine swap began in August. The crew of mechanics and engineers had a delicate operation to carry out: a surgery in aluminum and magnesium. In a stroke of luck, the Allison and Merlin engines were just about the same size—the distance from the firewall, the rear mounting for the engine, to the forward engine "mount" was almost identical, though the Merlin was 290 pounds heavier.

Like a Jarvik-7 artificial heart, the Merlin 61 now had to be implanted into the P-51A body. The engine, 1,640 pounds and just over seven feet long, fit into the cowling, or nose, without creating any bulges or blisters. But suspending and mounting the big dynamo was a problem. The source of the glitch was the size of the superchargers: the housing for the Merlin superchargers was too wide to be secured in the engine bay of the P-51A. Fixes must be made in the prow of the plane.

External changes to the body of the Mustang were also necessary. The intercooler radiator was moved to a position under the engine. An old air intake was removed from the top of the prow because it interfered with the pilot's forward vision. The engine was elevated 3.5 inches. The fuel system was found unsuitable for high altitudes up to 41,000 feet: a new pumping system was created. Finally, a great spanning, four-bladed, eleven-foot-four-inch wooden Hydrolignum propeller made by the Rotol company was mounted on the spinner of the new creation, replacing the three-bladed prop: a bigger propeller to match the more powerful engine.

In early October the first converted plane, serial number AL975-G—G signifying that the plane must be guarded at all times when on the ground—rolled out for ground testing. The British dubbed it the Mustang Mark 10, or X.

On October 13, 1942, Rolls-Royce's chief test pilot, Captain R. T. Shepherd, took AL975-G, the first Merlin Mustang in history, into the air. In all he made seven test flights. And then, on November 13, he made his definitive sortie. It was high noon for his wings. Speeding across the Nottingham skies, racing north to Sheffield, south to Lincoln, diving, turning, the plane reached a top speed of 427 miles per hour at 21,000 feet. As it flew higher and higher, the Mustang no longer lost speed—it gained it.

The newly minted Mustang X had performed like a genie out of a lamp. The results were stunning. The top cruising speeds of the

Spitfire and Hurricane—374 and 322 mph—could not begin to compare with the Mustang's. And testing of the captured Fw 190 had revealed a maximum speed of 405 mph at 20,000 feet compared to the Mustang's 427. With its new engine, the Merlin Mustang could outfly anything in the Allied—or the Luftwaffe—inventory. London had its answer to the Germans.

There was general rejoicing. Rolls-Royce threw a dinner party for everyone who had worked on the conversion project. By all accounts it was less than festive. The war-rationed food was served almost cold, and in the blackout one guest never made it to the party. In the dark, he stumbled down a hole being dug for the foundations of a new shed and broke his arm. But legend has it that during the dinner Hives and Colonel J. J. Llewellin, the minister of aircraft production, laid a wager: Who would be the first in the air with a converted Merlin Mustang, the British or the Americans?

Six thousand miles away, in California, Edgar Schmued had been following events in England. Through the spring of 1942, he had kept track of the British conversion project; soon he began an engine swap of his own. Ten days after the first Mustangs were delivered to Rolls-Royce, North American began work at Mines Field on two P-51As assigned by the U.S. Army Air Forces for a switch. North American now joined the race to graft a British Merlin engine into an American Mustang airframe.

The first problem facing Schmued was performance. The Mustang had maneuvered faultlessly with the less powerful Allison engine on board. But once it had the new Merlin 61 implanted, Schmued, as the designer, knew the plane would not handle well. It would be more difficult to fly. Ever the perfectionist, he insisted that his team find a way to use the new engine without losing the aerodynamics of his original design. Drag and lift—the old gospel of Gustave Eiffel, Otto Lilienthal and the Wright brothers—were still the key. Designers, draftsmen and engineers now began work on the aircraft—air intake, engine cowling, the propeller, a new carburetor—reshaping the sleek avian that had first come to life in their shop two years before. The British, far off in Europe, were racing against the German Focke-Wulfs with the Nazi threat at their backs. But the crew in the hangars in Inglewood, California, worked with no less urgency.

Between August and November 1942, North American put in 223,000 man-hours reconfiguring the plane and modifying the airframe. A new magneto had to be found, a different air cooler. Changes had to be made to the water pumps, the superchargers, the water-alcohol injection system. As the months lengthened through autumn, Schmued pushed his men hard to complete the job.

Finally, the holdup was the Merlin engine itself. The version Schmued would use was an American model being built on license from Rolls-Royce by the Packard Motor Car Company in Detroit. Like Rolls-Royce, Packard was a luxury car company that had turned to making aerial engines in World War I. Rolls-Royce had realized as war threatened again that it would need far more Merlins than its own plants in England could churn out. They had approached the Ford Motor Company about manufacturing the Merlin in the U.S. under license. But Henry Ford, an isolationist who believed Britain would surrender to the Germans, rejected the overture, and in September 1940 Packard signed an agreement with Rolls-Royce, worth $2.5 billion in today's dollars. By war's end Packard would make more than 55,000 Merlins.

But there were problems: manufacturing methods at Packard were very different from those at Rolls-Royce, and the Americans in Detroit needed time to tool up. Rolls-Royce still made engines by hand, with workers learning a new job through verbal instruction rather than detailed drawings. Packard used an assembly line and mass-production methods. Another sizable hurdle was that British and American screws were different: Packard would need to manufacture tens of thousands of screws and bolts never used in the U.S. before.

The British-American collaboration on the Merlin was unique. No weapon in the Allied arsenal, from the M1 carbine rifle to the *Essex*-class aircraft carrier, had been developed by two nations in partnership during the war—with the possible exception of radar. And the Merlin engine itself was mind-bogglingly complex.

The Merlin 61 had 14,000 separate parts. There were eighteen specific differences between the Packard V1650-3, the American version of the Merlin 61, and the original Rolls-Royce engine. To get the job done, a blizzard of 9,000 drawings would be exchanged by the two companies. Rolls-Royce would send three of its engineers to Detroit, one to remain for the duration of the war. Production of the Packard Merlin V1650-3 would lag behind the conversion at North American—and the completed airframe would sit for weeks on the tarmac at Inglewood waiting for its engine. Colonel Llewellin would win his bet: the Americans would be more than a month behind the British getting their plane into the air.

On November 30, 1942, Bob Chilton, a test pilot for North American, flew the P-51B—B designating the new engine—for the first time. The flight was no home run. Chilton flew for barely 45 minutes before the Merlin engine overheated and seized up. Baffled, the engineers in Schmued's shop sought out the Bureau of Standards in Washington for advice. The solution that came back was almost laughably simple: coat the radiator with Keg-Liner, a lacquer used on the insides of beer cans to isolate the beer from the metal container.

On December 4, with the overheating problem solved, Chilton resumed testing. He took off, climbed through the clouds, banked, barrel-rolled, soared and dove over Southern California, putting the plane through its paces. This time it performed perfectly—and a bold new breed came alive under his hands.

With the X designation—X for "experimental"—no longer needed, the XP-51B became the P-51B. Faster than any rival in the air, with the speed of a thoroughbred and the endurance of a plow horse, it seemed poised to charge into the flak and fire of the war in Europe.

But on the cusp of victory, the inspired fighter would now enter a wasteland of delay and neglect. Somehow, no one on the American side of the Atlantic seemed to be in a hurry to get things going. The plane was assigned a dismal priority by the U.S. Army Air Forces, and production did not move forward. Then, in December, the Mustang got lost. In England, those who knew what it could do could not understand why.

PART TWO

BIRD IN A CAGE

Tommy Hitchcock

CHAPTER 8

The Best Sport in the World

Almost exactly one year earlier, on Sunday, December 7, 1941, Winston Churchill sat at dinner with Gil Winant, the American ambassador to Great Britain, and several others at Chequers, his official residence in the countryside 100 miles outside London. The British prime minister asked Sawyers, his butler, to put a radio on the table. It was a small $15 receiver that Harry Hopkins, President Roosevelt's closest advisor, had given to Churchill. Static. The crackle and hiss of a news broadcast came on. They had missed the first moments of the evening report.

Then the lead bulletin was repeated.

The Japanese had attacked the U.S. Pacific Fleet at Naval Station Pearl Harbor in Hawaii that morning in a surprise assault.

The listeners were incredulous. The impetuous Churchill, who had prayed for America's entry into the war, jumped to his feet, announcing he would immediately declare war on Japan.

Reconsidering, he said: "I will call up the president by telephone and ask him what the facts are." He added: "And I shall talk with him, too."

The call went through in minutes. FDR gave Winant a stark summary of the attack, though he could not go into details on an open telephone line. Churchill took the phone. "It's quite true," FDR told him. "They have attacked us at Pearl Harbor. We are all in the same boat now." It was 11:00 a.m. in Hawaii.

Just after 7:50 that morning, wave upon wave of torpedo planes and dive-bombers, launched from an assault fleet of six Japanese aircraft carriers, had screamed in over the U.S. Pacific Fleet anchorage. The 353 attack aircraft flew in low over the oil tank farms, shipyards, buildings and docks of Pearl Harbor as the Japanese sank the battleships *Oklahoma* (BB-37) and *West Virginia* (BB-48), and then in quick succession the *Arizona* (BB-39) and the *California* (BB-44). Numerous other ships were damaged, including the heavy cruisers *New Orleans* (CA-32) and *Raleigh* (CL-7) and the destroyers *Helm* and *Hull*. In all, 2,335 servicemen were killed or missing in the strike; 1,143 were wounded, in addition to 103 civilian casualties. Fortunately, the American aircraft carriers were at sea and the Japanese failed to destroy shore and oil-storage facilities.

Only 15 minutes before the attack, Japanese diplomats were negotiating for peace; their declaration of war arrived in Washington as the strike at Pearl was underway. The next day, shortly before noon, a long black limousine carrying President Roosevelt, dressed in high silk hat and black naval cape, left the east gate of the White House and made its way slowly to the Capitol. Through the side door leading into the House of Representatives stepped the president's son James in his blue Marine uniform. Lurching from side to side, one arm locked in his son's, Roosevelt gripped the rail with the other and inched up to the dais. He opened his loose-leaf notebook, waved his arm and adjusted his eyeglasses. Then, in his familiar tenor voice, he asked Congress to declare war on Japan.

"The days of peace were over for us," Ambassador Winant would write in his memoir. Three days later, the United States was also at war with Germany and Italy.

In Eastern Europe, Nazi fury continued to burn. In the Soviet Union that June, the Germans had launched their Blitzkrieg duo of Panzers and Luftwaffe across 1,800 miles to the outskirts of Moscow. Hitler believed he could subdue Stalin's forces in a few months. But in December the frost halted the German onslaught and the Russians began to dig in to stop the Thousand-Year Reich. Two days before Pearl Harbor, a massive Russian counteroffensive was launched. The VVS, the Soviet Air Force, and its Ilyushin IL-2 Shturmovik attack planes flew 51,300 sorties in defense of Moscow.

It was the worst winter in 20 years, and it would prove to be the coldest winter of the twentieth century in Europe. Temperatures plunged to −40 degrees Celsius in some parts of Russia. In the intense cold, a mere 15 percent of the Luftwaffe's 3,000 aircraft were operational. German planes could be started only by the most desperate means, sometimes by building fires under their engines. Mechanics' hands froze around their tools and had to be loosened with heaters. German infantry fought in constant freezing temperatures, enduring snow and ice as food supplies dwindled. By October and November, frostbite cases had decimated the ill-clad Germans, for whom there was insufficient winter clothing, while the icy cold paralyzed the German mechanized transports, tanks and artillery. In the end, 2.5 million men and 162 divisions advanced almost in futility, slowly halted and then were lost in the freezing immensity of the Russian winter. Hitler's armies were reaching their limit.

But the lair and lodge of the Nazi war machine, the citadel of arms production from which Hitler's war—in Africa, the Mediterranean,

Russia—was marshaled and sent out, was itself untouched. Germany continued to mill out huge numbers of arms, remaining an impregnable rampart of military production. With Germany all but landlocked, attack from the sea was not possible; with the Third Reich holding all the sprawling extent of Fortress Europe, armies were hamstrung.

Only the Allied bombers would be able to strike at the centers of Nazi arms production. Thus the urgent clash would soon depend on the air war; the air war would rest on the bomber war; and the bomber war itself would ultimately turn on the Allies fielding a long-range fighter aircraft with the endurance to protect the bombers journeying deep into the battlements of the German Reich.

On paper, Lieutenant Colonel Tommy Hitchcock—not Thomas, not Tom, but always Tommy—did not seem a likely candidate for the job of finding the Allies' doomsday weapon. Fair-haired, handsome, with the nerve of a daredevil and the personal code of a gentleman, he was a man of many parts, but at first glance none of them would explain the obsession he developed with the Mustang.

With roots on both sides extending back to American families prominent since before the Civil War, Tommy had grown up both in the horsey oasis of Aiken, South Carolina, and on Long Island's affluent North Shore. His education included the elite boarding school St. Paul's, Harvard and Oxford. A son of privilege and inherited wealth, he would go on to marry an heiress, Margaret "Peggy" Mellon, the widowed daughter of Pittsburgh financier and Gulf Oil founder William Larimer Mellon, and make a successful career in investment banking as a partner at the patrician Wall Street firm of Lehman Brothers.

Tommy trod the lawns of the well-heeled and stylish; he was even a close friend of the writer F. Scott Fitzgerald. Indeed, his astral place in New York society was at least part of what attracted Fitzgerald, who met Tommy at decadent parties on the North Shore in the 1920s. These were the Roaring Twenties, when "in speakeasies and private houses, on lawns and beaches, small parties and enormous parties" were a way of life for the rich. Fitzgerald idolized Tommy, aspiring to belong to the American aristocracy Tommy had been born into. He modeled two of the protagonists in his novels on Hitchcock: Tom Buchanan in *The Great Gatsby* and Tommy Barban in *Tender Is the Night*. The Hitchcock compound at Sands Point was just across Manhasset Sound from the Fitzgeralds' house and became the fictional town of East Egg in *Gatsby*. The Fitzgeralds were regulars at Sands Point, often becoming the guests who would not leave. "He would sit in the house," Hitchcock's eldest daughter, Louise Stephaich, remembered, "and they didn't know how to get him out. He had this crush on my father. He wanted to be like my father. He was mesmerized."

Tommy moved beyond this glittering social scene to become internationally famous as a world-class athlete. In the twenties and thirties, Hitchcock became a polo star—but he did not just play polo; he reordered the history of the game. Having led the U.S. team to mastery in the 1921 International Polo Cup, he had a ten-goal handicap from 1922 to 1940, the highest ranking in the sport. As a player he was fierce, single-minded and aggressive. José Reynal of the Argentinian team recalls watching Tommy smack a ball 170 yards downfield. Then at midfield Hitchcock overrode the ball at breakneck speed, checked his horse hard, turned and, with Reynal bearing down, slapped off a quick shot. "I sat helpless and watched the ball climb into the wind," Reynal recalled in an article by *Sports Illustrated*. "It wavered to right and left but held a generally true line. Finally it came to earth and rolled between the posts. It didn't seem possible."

By the 1930s, Hitchcock had become a symbol of the sport: he was to polo what Babe Ruth was to baseball, Don Budge to tennis or Bobby Jones to golf. By the time he was 40, many said he was the greatest player in the history of the game. He was responsible for popularizing polo, making it a mass-spectator draw like football or racing. Stands were built for 40,000 spectators to see him play; even the Prince of Wales attended his games. When he arrived at a match, "a camel hair coat draped around his shoulders, crowds surged around him and applauded wildly," one writer observed. His biographer relates that Tommy liked to joke that "his own celebrity had gone so far in Hollywood that he had been approached by one of the companies to make a movie playing himself."

This gilded superhero of sport was the same man who, as a military officer in 1942, would discover the P-51 Mustang in its exile, seize its potential and possibility, and lead the vanguard that brought the plane to its pivotal role in 1944 in turning the tables of the war. And Tommy was a man of other contradictions, too: his powerful build, "wide, thick shoulders and big arms" belied the "gentleness of his voice and manner, his simplicity, his consideration." One friend called Tommy the only perfect man he had ever met.

Tommy was many things to many people, but his passion had always been aviation. As a student at St. Paul's School during World War I, he lived in an atmosphere charged with patriotism. By the time he turned 17, Tommy had caught the bug and was determined to fly in the war. The group that called out to him was the illustrious Lafayette Escadrille of the French Air Service.

In Europe, the mass slaughter in the trenches had gutted morale. Battle in the skies provided a cleaner form of warfare compared to the gruesome losses in No Man's Land. And fighter pilots offered welcome heroes to boost public feeling about battle—though in fact the air war had more in common with the war on the ground than people thought. Air combat was ruthless, and British pilots arriving at the front in 1917 had an average life expectancy of two weeks. But fighter pilots who became aces were instant superstars.

The French were the first to create elite formations of fighter pilots, called the *Cigognes* (Storks). In 1916, to encourage the Americans to enter the war on the Allied side, France created another elite group of pilots, the Lafayette Escadrille—named for the Marquis de Lafayette, a French hero of the American Revolutionary War—from American volunteers. It became the most famous fighter group in history. Stories of the Lafayette stirred the imagination of everyone at St Paul's and appealed to the families of the elite like the Hitchcocks.

At 17, Tommy dropped out of St. Paul's to join the Lafayette Flying Corps, the successor to the Escadrille. Because he was underage, Theodore Roosevelt, an old family friend, was asked to pull the necessary strings. By late spring, the adolescent Tommy was on his way to France, the youngest American to be brevetted by the French *Aéronautique Militaire* in the First World War. He trained at the *École Militaire* at Avord, the best and largest flying school in the world. In October 1917, he graduated as a pilot and corporal.

Joining Squadron N-87 of the Lafayette Corps, in the blink of an eye Tommy was rocketed away into the pulse of a fighter pilot's life. Soon he would fly against the very planes the young Edgar Schmued, exactly the same age, was repairing and servicing for the Austro-Hungarian Air Service. His first days were cold and damp—the area was socked in with clouds, poor weather for flying. In the hangar Tommy would down a breakfast of bread and chocolate, chatting with his mechanic "Noodles" and his fellow pilots, including William Wellman, the future Hollywood film director. In January, Tommy baptized his sleek little Nieuport biplane: on his first reconnaissance patrol, he flew too high and found himself alone in the sky. When he saw a German two-seater—the ultimate prize—he dove straight on the plane and started shooting. The Boche trailed down through the sky, spun through the clouds and crashed nose down on a hillside in the Vosges Mountains. Corporal Hitchcock, one month shy of 18, had a kill and his indoctrination in combat. He celebrated that night with a bath and a victory dinner at the Grand Hotel de la Pomme d'Or. From there, he received the Croix de Guerre with Palme, was assigned a newer, better Nieuport and on January 19 was aloft again.

Above in the gypsy sky once more, he soon saw his wingman peel away with engine trouble. Not long after that he spotted it—another fat two-seater. He plunged down, swooping in from the sun, ambushing the German plane and firing off a good burst. The Boche spun stricken, plummeting toward the ground, winding and spiraling until it crashed into a field. As Tommy roared over the two-man crew, the Germans waved up at him. Hitchcock had been flying only two months and he already had two victories. "I live for the flying," he wrote to his uncle, "and the flying alone."

In March, Tommy's luck ran out. Patrolling over the north of France near Nancy in good flying weather, the young airman spotted two German D.III Albatroses. Splitting off from his squadron, he dove on them, not seeing in the sunlight above a third German trailing his plunging Nieuport. Hitchcock heard the sound of gunfire and blacked out. Seconds passed. The plane veered. He blinked awake to see the earth whirling wildly beneath him. Frantically he regained control. As he headed back toward western France, a fearsome burst of tracers passed between the right wings of his biplane and he found himself staring back into the face of the German pilot.

Another burst of tracers and Hitchcock heard something give

way on the back of his plane. He stamped hard on the left rudder and realized his right leg had no sensation. The gallant little Nieuport, almost flying itself now, came down in a field and crashed, breaking off a wing. Barely conscious, Hitchcock saw German soldiers run out of the woods. He blacked out.

He was treated at once at a German first-aid post, fainting again from the pain. He had been lucky: the bullet he had taken through his hip had not hit bone. But the wound became infected and did not heal. From there the months turned into a constant shuttle as Tommy was transferred between German hospitals and prisons by every mode conceivable: trolley, oxcart, train, sometimes even walking in excruciating pain. He thought ceaselessly of escape. Landshut. Lechfeld. By train to Rastadt. As the train shoved slowly through the Stuttgart countryside, Tommy brooded, silent and watchful. He figured the guard must fall asleep at some point, and the man had two tantalizing possessions: a train schedule and a large map of Germany.

"All I had to do was to locate the map . . . and transfer it to my coat pocket," he later wrote. "Still I was scared to death that, as soon as I got the map out in the open, the guard would wake up and all would be lost. . . . I made several feeble attempts . . . as if my hand were unconsciously moving in my sleep." Then he had it; he had the map and got it in his pocket. There was no time. Such an opportunity would not come again. At once he bolted for the door, flung it aside and jumped into the darkness. He stumbled over a hedge, raced across a lawn and plunged into a thicket. As he stopped to catch his breath, he heard the train pull away. "I was free. Free to turn this way or that, free to go or stay still. What a joy to be walking about once again in the open country," he recalled. He was just 18 years old.

Tommy's brush with death was not enough to dampen his passion for flight, even after the Great War ended. In the thirties he took up flying again: he bought a seaplane and, in the summer and fall, flew it in "from Sands Point to the East River in the morning and back out again in the evening." His children, sometimes along for the ride, were thrilled by his daredevil stunts in the air. As war once again threatened in Europe, Tommy's love of soaring found a new expression: he created the American Export Airline Company with passenger and freight service to Lisbon by transatlantic flying boat.

In the thirties flying boats "enjoyed a brief golden age . . . as the only aircraft offering passengers transportation across the oceans." Travelers viewed them as safer because they could land on water. Their boatlike hulls could be furnished with luxurious accommodations: "spacious compartments, upholstered chairs, backgammon tables and hot meals served by a uniformed steward." This was true luxury travel; airfares could run into many thousands of dollars.

Tommy was intrigued by this new frontier. By the time he started his aviation company in 1937, his fierce competitor for flying boat passengers was also a friend from New York high society: Juan Trippe, the illustrious president of Pan American Airways. Pan Am already had a monopoly on flying boat travel, transporting passengers in Sikorsky S-40s, the first Pan Am planes to be called Clippers, up and down the Central and South American coasts. When war broke out, Hitchcock and Trippe were still locked in competition for landing rights for their flying boats.

Until Pearl Harbor Tommy had been an isolationist, an unpopular position among his contemporaries. One of the reasons was surely his experience in the First World War. It had left him disgusted with the ineptitude of governments that had squandered needlessly so many millions in battle, equipping his fellow young pilots, for example, with obsolete airplanes that took them to their deaths. The planes flown by the Lafayette Escadrille had been flimsy and unsafe: engines failed and machine guns often jammed. Wounded seriously in combat himself, he shrank from the thought that his country should be plunged into it again so soon. But once war was declared, his only thought was to sign up and serve again as a pilot. "Polo is exciting," he would say later. "But you can't compare it to flying in wartime. That's the best sport in the world."

The day after Pearl Harbor Tommy picked up the phone. His biographer, Nelson Aldrich, wrote that he knew more people than God, and now he milked his contacts to try to get into the war. Aldrich amiably observed that, "Socially, wartime Washington was like a Harvard-Yale football game, except bigger, more jovial" and Tommy knew all the players. In Washington he paid a call on Major General Henry H. "Hap" Arnold, chief of the Army Air Forces. He stopped in at the Pentagon and called on officers at Langley Field. He entertained air force brass at his hotel, pressing everyone he saw with the same question: how could he get into the war as a pilot? And everywhere the response was the same: he was over 40 and too old to fly. Further, he had no experience flying the new monoplanes that would now dominate the war—the P-40s, P-38s and P-47s.

Finally, it was the old-boy network, from his St. Paul's School days, that swept Tommy up and projected him into the midst of the newly kindled war. At lunch with his wife, Peggy, at the Shoreham Hotel in Washington, he ran into Gil Winant, the ambassador to Great Britain, back in the States for consultation. Winant was an old friend. He had been Hitchcock's teacher and mentor at St. Paul's and the master of his residential house, New Upper. In his school days, they had sat up many nights talking: Tommy about his dreams for the future, Winant about his ambition to run for the New Hampshire State Assembly. Both had flown in combat in World War I, Winant with the 8th Aero Squadron, a U.S. Air Corps unit that fought on the Western Front. He understood Tommy's disappointment: he would create a job for him as assistant air attaché at the London embassy, serving as a link and liaison to Britain's Royal Air Force and its Fighter Command.

By late April, Tommy was on his way to England, leaving Peggy and the children behind in their elegant apartment at 10 Gracie Square in New York City. If he could not fly, he could at least be close to the developing air war in Europe.

CHAPTER 9

A One-Man Crusade

Tommy arrived in a London sooty and grim with the wreckage of war. Empty lots of jagged rubble where houses had once stood served as a bleak reminder of German bombers that had ravaged the skies only a year earlier. At night the city lights were blacked out; armed guards carrying shotguns patrolled the streets. The furious onslaught of the Battle of Britain and the Blitz were gone, but in their place was a pallid passage of endurance. As the novelist Elizabeth Bowen wrote of that time in 1942, "It was now, when you no longer saw, heard, smelled war, that a deadening acclimatization to it began to set in. . . . This was the lightless middle of the tunnel."

The rationing was doleful. Londoners were restricted to a bare two ounces of butter per week, three pints of milk and one egg; restaurant meals were limited to five shillings, or about $1.25. Many neighborhoods were gutted and left with open spaces where buildings had once stood. Everywhere dust, rubble and construction debris cluttered the streets and blocks of the capital.

Tommy walked the streets and squares of this gray netherworld glum and forlorn. "London is dreary," he wrote to Peggy from temporary quarters at the exclusive Claridge's Hotel. "There are shortages of all kinds, particularly amongst the civilian population. The town is blacked out at night, plays go on at 6:30 to 7 so the people can get home by daylight. The signs of last year's bombings are . . . left where the houses were. It must have been pretty bad when it was going on."

Two years into the hostilities, Londoners seemed stoic, going about their business with matter-of-fact calm. This only accentuated his loneliness. "When I wake up here I wonder what the hell I am doing in a strange city miles away from everyone I love and care about . . ." he wrote to Peggy that June. At 42 and assigned to a desk job, he felt he was sitting out the war, pushing paper. He tried to fill his time dispatching report after report to Washington on British planes being tested by the RAF as America, a latecomer to the war, accelerated its aircraft development program. But within weeks of arriving in England, Tommy had found his mission.

Ronnie Harker had just made his first flight in the Mustang and filed his historic report. The P-51A was being tested at the RAF's Air Fighter Development Unit in Duxford; hearing of it, Hitchcock set about learning all he could. Calling at Duxford, he observed performance tests of the new fighter and, according to his biographer, "spent long hours poring over reams of charts, graphs and statistics comparing the performance of the Mustang to the Spitfire IX." The numbers clearly showed that the Mustang, though 1,500 pounds heavier than the Spitfire, was indeed 30 miles per hour faster at lower horsepower. It was just as Challier had predicted. The more Tommy learned, the more excited he became.

By the end of May, Tommy's faith in the avian had become a creed. In a report for Army Air Forces headquarters in Washington, he recommended the Allison Mustang be fitted in America with the Merlin engine, then being used in the Spitfire. This, of course, was precisely the project Rolls-Royce was about to initiate. "It is estimated," he wrote, "by the RR people that the Mustang equipped with the 61 engine will have a top speed of 440 miles per hour and will have a speed curve of somewhat more than 20 miles per hour faster than the Spitfire IX." He predicted the conversion "would produce the best fighter plane on the Western front."

By June, Tommy had flown a P-51A himself. "Yesterday I had a good flight on a new fighter the English call the Mustang," he wrote to Peggy. "It is made in California by the North American Co. This is a great airplane and very fast." He was brushing up his World War I flying skills. "I am getting a little more confidence in the high-performance stuff," he wrote, ". . . at least I can fly without my knees knocking together."

In no time at all Tommy's infatuation with the Allison Mustang had traveled to the well of his heart. All that summer and fall, he visited airfields scattered across Britain in the embassy's Beechcraft plane, gathering data and talking with supporters of the Mustang conversion at Rolls-Royce and in the RAF. Often, he found himself at the Rolls-Royce Experimental Installation Plant near Nottingham, where the transplant operation was about to begin on four Mustangs. More and more, among the whole lineup of Allied fighters, he was coming to see the Merlin Mustang as a silver arrow, the essential weapon in the rapidly developing air war.

Like Ronnie Harker, Tommy perceived, with a good deal of prescience, that the Allies needed two qualities in a fighter plane: enough performance to outfly any of the Nazis' spectacular aircraft, and as the war entered a new phase, enough range to shield bombers deep inside Germany. Like Harker, he thought the Mustang refitted with a Rolls-Royce engine would be the Allies' ace in the hole.

The British were already convinced. Now the most important American hat, U.S. ambassador Winant, joined the Merlin Mustang crusade.

Winant was a three-time governor of New Hampshire and a friend of FDR's; his gaunt frame and dark deep-set eyes gave him a strong resemblance to Abraham Lincoln. Tall and lanky, a shy man of ungainly charm, Winant had arrived in Britain in March 1941 to succeed the unpopular isolationist and Hitler appeaser Joseph P. Kennedy as the Blitz still raged. The British had sent a special train to western England to meet the new American ambassador. From the moment King George VI personally met him at the railway station and entertained him for tea at Windsor Castle, Winant was seen as a friend to the British. A frequent visitor to 10 Downing Street and to Churchill's residence at Chequers, Winant forged a close relationship with the prime minister. Churchill's private secretary, John Colville, called him "a gentle, dreamy idealist, who most men and all women loved."

Winant was a pilot at heart and always took a personal interest in air strategy. Infected by Tommy's enthusiasm, he quickly became a true believer in the Mustang; he immediately understood the importance of its unique capability. He and Tommy began to meet with air marshals Charles Portal, Sholto Douglas and Francis Linnell and American general Alfred J. Lyon, chief technical officer for the USAAF in Britain, as well as senior staff from Rolls-Royce and North American Aviation at the center of the rapidly growing Mustang vortex. With Tommy leading the charge, the two men now lobbied Washington together from the American embassy on behalf of the airplane.

The embassy in Grosvenor Square had been much transformed by the war. By 1942, the eighteenth-century Georgian residences and gardens had been torn down and replaced by blocks of luxury office and apartment buildings, many of them damaged by German bombs. The neighborhood teemed with American soldiers and airmen, as well as U.S. government officers and dispatch riders racing between the embassy and the British War Office. At his apartment on the upper floors of the embassy building—with offices underneath and staff rushing back and forth around the clock carrying Army and Navy signals in code—Winant became Hitchcock's partner in the crusade for the Merlin Mustang. There, sitting together, the ambassador and Tommy plotted out their campaign to sell the plane in America, plying FDR and the Pentagon with cables, letters and reports on the performance of the new bird of prey. Winant worked closely with Churchill on telegrams that were sent to Roosevelt urging action on the P-51. He personally contacted high-ranking American officers arriving in Britain to press the cause of the Mustang, making the same case at the Pentagon on his visits to Washington. He and Tommy relentlessly pushed everyone who might help move the project forward.

But soon it became apparent that the U.S. Army Air Forces, invested in American engines and American planes, were not on board. Inexplicably to the English and to Winant and Hitchcock, the scheme foundered. It was then that Hitchcock decided that, whatever the fate of the Merlin Mustang in America, he would make certain it came to the attention of the American generals running the air war in England. Tommy was determined to introduce British supporters of the Mustang—at Rolls-Royce and in the RAF—to the generals of the U.S. Eighth Air Force, then just starting up across East Anglia and swelling rapidly to maturity.

The Eighth Air Force was growing from a seed to a spreading oak. When General Ira Eaker and six staff members arrived in England in February 1942 to create the new force, they were starting from scratch. At once, what would become the "Mighty Eighth" Air Force shot up like a sapling. By April, Eaker had taken over Wycombe Abbey, a girls' school, as the Eighth's headquarters. Code-named "Pinetree," it was located on elegant grounds an hour northwest of London, near Southdown, the headquarters of RAF Bomber Command. Billeted at first with Air Marshal Sir Arthur Harris, the controversial head of Bomber Command, and his family, Eaker struck up a fast friendship with his British counterpart. In some ways Eaker and Harris were unlikely friends. Their backgrounds were not similar: Harris was a former gold miner and tobacco planter from Rhodesia; Eaker, from rural Texas, was the son of sharecroppers. Harris liked his cocktails; Eaker was abstemious. But their friendship flowered despite their different personalities.

Harris could not have been more generous to Eaker: he turned over five established air bases in East Anglia to the Americans, with the British Air Ministry helping to build another 60. They provided the Americans with "food, clerical help, office space, motor vehicles . . . maps, projectors . . . photo equipment, miles of telephone wire and endless bits and pieces such as escape kits." With Harris's help, the British Isles would become the largest stationary aircraft carrier in human history, an immense launching stage for Allied bomber crews and the fighter pilots supporting them. From nothing, Eaker and his staff would build the Eighth Air Force to a vast array of 185,000 men and 4,000 planes by December 1943.

All of this transpired under the command of chief of the Army Air Forces General Henry H. "Hap" Arnold in Washington.

FIGHTER GROUP STATIONS ⭐
BOMBER GROUP STATIONS
FIRST AIR DIVISION STATIONS ▲
SECOND AIR DIVISION STATIONS ●
THIRD AIR DIVISION STATIONS ■

North Sea

56th FG-Horsham St. Faith
352nd FG-Bodney
20th FG-King's Cliffe
359th FG-East Wretham
364th FG-Honington
361st FG-Bottisham
358th FG-Leiston
Northampton
Cambridge
339th FG-Fowlmere
78th FG-Duxford
479th FG-Wattisham
355th FG-Steeple Morden
361st FG-Little Walden
356th FG-Martlesham Heath
4th FG-Debden
353rd FG-Raydon
Felixstowe
55th FG-Wormingford
354th FG-Boxted
Oxford

LONDON

U.S. Eighth Air Force
Installations in Britain
1944
25 KILOMETERS

English
Channel

Arnold is today considered the father of the modern U.S. Air Force. As a young army officer, he had learned to fly from the Wright brothers themselves, having enrolled at their aviation school in Huffman Prairie, Ohio, in 1911, paying $250 for a two-week course. "On Sundays, the Wrights would invite [Arnold and Tommy Milling, a friend from West Point] to dinner, and after stuffing themselves with food, the young men would sit transfixed, listening to the two brothers tell their stories," recounts one of Arnold's biographers. During the First World War, Arnold helped expand the Army Air Service, then in its infancy. (Throughout World War I, the Air Service conducted only 150 bombing attacks and, with 740 aircraft, comprised only one-tenth of Allied airpower.)

Following the armistice ending World War I, Arnold returned stateside, the fighting ceasing before he could see combat. In 1919, he took command of Rockwell Field near San Diego, with the blunt, slightly built Carl A. "Tooey" Spaatz as his executive officer. He recruited soft-spoken, square-jawed Ira Eaker as his adjutant. Together, the three men would embark on a crusade to build a true American air force.

Arnold, Spaatz and Eaker were all pioneers of the air in their own right. In 1929, Spaatz and Eaker stayed in the air for a record 150 hours using midair refueling. Arnold led a flight of bombers on a nonstop flight from Washington State to Alaska over trackless mountains without an air route in 1934. And in 1936, Eaker made the first transcontinental flight on instruments alone.

In 1932, Arnold became skipper of March Field—near Riverside, California—where Spaatz had command of a bomber wing and Eaker a pursuit squadron. Finally, in the late thirties, he became chief of the Air Corps, the successor to the Air Service, and began to make his plans to lead the emerging American aerial force. Spaatz, at his side for 20 years, was there to shape strategy. Eaker was close at hand, too. As soon as war was declared, Arnold would send Eaker to England to prepare the groundwork for the American bombing assault in Europe while Arnold in Washington, like a snake charmer, would call forth from his basket the pilots, ground crews, support staff, fighters and bombers that would grow into the Eighth Air Force.

By the time Tommy arrived in the spring of 1942, the Eighth had grown to a muster roll of 1,871 men, all ground staff with no planes. Pilots now began to fly new bombers across the Atlantic; fighters were flown to the East Coast, disassembled, packed and sent across on shipboard or flown across to bases in Britain. Airfields were built and paved; barracks, mess halls, mail posts and weather stations sprouted up. In June, the passenger liner *Queen Elizabeth* brought aircrews of the 97th Bomb Group; made up of three squadrons, it was the first to arrive. Pilots and navigators followed within a month. Flying Fortresses now flew in numbers across the Atlantic to England. The germ of an idea had become an air force.

The U.S. Army Air Forces, of course, were not alone in this war. England's Royal Air Force flew in defiance of Nazi Germany long before the Americans took flight and was the indispensable partner of U.S. forces throughout the air war in Europe and Asia. The RAF was the first independent air service in history, created on April 1, 1918; it had grown from 42 squadrons and 800 aircraft in 1934 to 157 squadrons and 3,700 planes by the war's outbreak in 1939. With its dual mission of defending Britain and fighting the world war against the botulin of Hitler, it waged two distinct campaigns during World War II: the fighter campaign of RAF Fighter Command, which included the Battle of Britain; and the strategic offensive campaign of RAF Bomber Command. The bomber war of the RAF centered around night "area bombing" of cities and population centers and vast 1,000-plane raids. These were the brainchildren of its chief air marshal, Ira Eaker's friend Arthur Harris, a confidant of Churchill whose bluff, hearty manner was at odds with his image as "Bomber Harris" in the press and "Butcher Harris" to many in the RAF. "They sowed the wind," Harris famously said, referring to the Germans, "and now they will reap the whirlwind." By the end of the war his area bombing would outrage many for the human cost of missions against civilians.

With time, the RAF developed sophisticated electronic attack and defense systems. Its contribution to the Allied victory in the air war in Europe was enormous, and critical on all fronts.

But the American air war was central; in 1944, the British air arm would be placed under the direct control of General Dwight D. Eisenhower, supreme commander of the Allied Expeditionary Force, in preparation for D-Day.

In the summer of 1942, Tommy Hitchcock began his charm offensive to win over the officers of the U.S. Eighth Air Force to the P-51. After a few days at Claridge's, he had taken a furnished flat not far from the embassy. Now it became, Aldrich wrote, "like a club room of the combined general staffs." There he threw dinner parties to bring together the commanding officers of the British and American air services, as well as visiting dignitaries from the White House and the Pentagon. He marshaled his contacts, assembling a coalition of supporters for the Mustang. His flat was a stopping-off point for Averell Harriman, the American statesman and diplomat; Jock Whitney, the multimillionaire, sportsman and publisher of the *New York Herald Tribune*; David K. E. Bruce, later ambassador to En-

gland; and a raft of Eighth Air Force commanders and bigwigs as
well as British RAF brass. A frequent visitor to the apartment was
his sister Helen's son, Averell. At 21, the handsome, dark-haired Avy
had volunteered for the RAF, and in the spring of 1942 he was flying Spitfires in one of the three Eagle Squadrons of young American pilots in Britain.

At the same time, the British were doing all they could to make
sure that American officers in England "could see first-hand the
merits of the Merlin Mustang." That summer of 1942, British air
marshal Sir Francis Linnell had a small contingent of U.S. Army
Air Forces officers posted to the air station at Boscombe Down,
where RAF trials of new or modified aircraft were performed, now
including the Mustang. Alongside Linnell's efforts, Air Chief Marshal Freeman arranged for two of the five Mustangs being reen-
gined at Rolls-Royce to be handed over to the U.S. Eighth Air
Force for trials and evaluation.

That summer, Tommy's anticipation mounted. With every weld
and knock in the hangars at Rolls-Royce, the converted Mustang
came to life. Now she was an aluminum patient in intensive care,
stripped down and disassembled, undergoing an industrial heart
transplant. Soon the new hybrid, the P-51B, or the Mustang Mark X
to the British, would have a powerhouse dynamo inside an aerodynamic fuselage as slick as an eel. She would be able to climb to
30,000 feet in 12.5 minutes, 2,400 feet per minute, 40 feet per second. At 32 feet, three inches long, with a commanding 37-foot
wingspan, the Mustang Mark X would now have a maximum
speed of 440 miles per hour—maneuverable as a Maserati and swift
as a whippet.

By the end of October, the Mustang conversions in England
were complete, and Tommy wanted to fly a converted Merlin Mus-

tang himself. His nephew Avy, himself by now a pilot with the 4th, the first American fighter group, accompanied Tommy to the Rolls-Royce field at Hucknall, where the new Mustangs were waiting. It was a foggy English day with poor visibility, and even on a clear day Clark knew that his uncle had a tendency to get lost in the air. "After he'd taken off," Clark remembered, "I said to the Rolls man half-jokingly, 'Well, that's the last time you'll see *that* airplane,' and the guy panicked. He called all the airfields in the area to watch out for a wandering P-51. Sure enough, we soon got word that a P-51 had landed at a field twenty miles away. I flew over and picked him up. He'd gotten lost, but he was pleased as punch with the plane."

That chilly fall, Hitchcock summarized his conclusions about the new Merlin Mustang in another memo to his Army Air Forces superiors in Washington: "The Mustang is one of the best, if not the best, fighter air frame that has been developed in the war up to date. It has no compressibility or flutter troubles, it is maneuverable at high speeds, has the most rapid roll of any plane except the Focke-Wulf 190, is easy to fly and has no nasty tricks. . . . [In] all respects, except rate of climb, the Mustang appear[s] to do the best against the [Fw] 190."

As the Merlin Mustang project moved forward, the American bombers were running into the first of their troubles. That October, General Eaker sent out 108 heavy sky arks in a raid over Lille, France, including more than 20 B-24 Liberators, accompanied by 400 fighter planes, all with a range short of the target. The lead planes were equipped with the Norden bombsight, an aiming device with which proponents said a pilot could hit a pickle barrel from 20,000 feet. Up until then, the Luftwaffe had been largely absent. But this time the German fighters showed up in force to defend the steel and engineering installations of the Compagnie de Fives-Lille and the locomotive and freight car works of the Atelier d'Hellemmes, both critical to German armament and transport. The giant B-17s and B-24 Liberators were badly mauled.

"USAAF losses were four destroyed, four seriously damaged and 42 needing repairs. . . . The bloodletting had begun," writes historian Paul Kennedy. The air war was moving ahead, but, from Lille on, the price would grow dramatically higher. Ultimately it would prove far more costly, with greater casualties, than any general or planner had expected. The Lille raid demonstrated for the first time to U.S. leaders how effective German fighter defenses could be.

The British knew full well that only America, with its large population and robust economy, could mass-produce the Merlin Mustang. But they were thoroughly discouraged about the prospects of getting it done. From Rolls-Royce and North American Aviation on up through RAF air chief marshal Freeman, the American embassy, Hitchcock and Winant, and the prime minister himself, all efforts at lobbying the Americans in Washington to give the Mustang priority—from letters to memos, cables and phone calls on secure diplomatic lines—had fallen flat. Overwhelmingly, reaction from America indicated the USAAF had no interest in the P-51B. The generals in Washington and at Wright Field in Dayton, Ohio, the proving grounds for the Army Air Forces, barely grunted at each new mention of the Mustang. It seemed they dozed through the Merlin Mustang's development in a comfortable pattern of patronage, favoritism and neglect, content to push along more familiar, American aircraft. By now conversions of the P-51 to the Merlin engine had also been completed at North American, but still the Army Air Forces would not move forward. There was little more the colony of Mustang advocates in Britain could do.

———

Tommy decided it was time to take his crusade to Washington. Ignoring proper channels, he would go straight to the top. He would make the case for mass production of the Merlin Mustang directly to the supreme commanding general of the U.S. Army Air Forces, Hap Arnold himself. In October he wrote to Peggy: "It looks very much as if I shall be able to get home for a short time before Christmas. The developments that I have been working on over here will require someone to go back to the US. . . . I cannot think of a better person to do this job than your husband."

Before Tommy left England in November, he did one more thing: he picked up the phone and called Chesley Peterson. Lieutenant Colonel Peterson was the most celebrated ace of the Eagle Squadrons, the fabled American fighter units who flew for England before the U.S. entered the war. A tall, wiry blond with deep-set blue eyes, he was boyish and friendly but ultimately all business when it came to fighter operations. He had fallen in love with airplanes early in life, when he hitched a ride with barnstormers who had landed in his father's alfalfa patch in Idaho. Among the first American pilots to join the RAF in 1940, he was an aggressive ace who strafed and dogfought his way through the skies over Europe and became the youngest Eagle Squadron commander at 21. By the time he joined the American 4th Fighter Group in September 1942, he was a celebrity on both sides of the Atlantic.

Now commanding officer of the 4th, Peterson had been ordered to proceed from Britain to Wright Field in Ohio to fly all the U.S. fighters and select the best one for the Eighth Air Force in Europe. "I hear you are heading for the States to pick out a fighter for the Eighth Air Force. I'd like to talk to you about the P-51," Hitchcock said into the phone that day. He said that at Farnborough, the RAF research station just outside London, there was an Allison Mustang whose engine had been replaced with a British Rolls-Royce Merlin. "I saw it one day and asked if I could fly it," Tommy said. "I was terribly impressed. . . . With the Merlin engine [it] would have the Spitfire IX beat. It would be a dream airplane."

He invited Peterson to come down to Farnborough before leaving England and fly it. Peterson was not disappointed. "[It] was a beautifully handling airplane," he would say later. Peterson saw that "[the] P-51 could be what it ended up being . . . the finest fighter that was built during the war." When he got to Wright Field, he determined, he would ask to fly another one.

CHAPTER 10

A Cross-Country Shuttle

Hitchcock arrived in Washington, DC, early on a windy November Wednesday, rumpled and travel weary, his fair hair disheveled by the breeze. The whole capital, from the winding Potomac River to the taxi lines outside massive Union Station, bustled with wartime activity. Still very much a small Southern town until Pearl Harbor, Washington had boomed as the country went to war. Stately federal buildings were now interspersed with Quonset huts; the city swarmed with new federal employees and heavy automobile traffic. That afternoon, at his sister's house, he prepared his papers, readying himself with tense determination for the presentation he would make to General Arnold.

The next morning, accompanied by Air Marshal Sir John Slessor of RAF Coastal Command and U.S. brigadier general Alfred J. Lyon, who had left his sickbed in England to attend the meeting (he would die ten days later), Hitchcock called on the chief of the Army Air Forces in his ring office at the Pentagon, the gigantic new five-sided military headquarters just completed in the mud flats alongside the Potomac. There Hitchcock, Slessor and Lyon made the urgent case for the Merlin Mustang. Papering Arnold's walls with graphs and charts, the three visitors bombarded him with data about the Mustang. They went over the clear speed and performance advantages of the plane, touting its long-range capabilities. They talked about the urgent need to escort the bombers deep inside Germany.

But the stubborn, irascible Arnold would not budge. The people at Materiel Command at Wright Field, the headquarters of USAAF arms procurement in Ohio, were completely satisfied with the Mustang's Allison engine, he grunted. It was, after all, an American engine. American engines were better. Why would we want a British one? The P-51B was no more than a fanciful experiment, Arnold went on, and a British experiment at that—not proven like the aircraft already in U.S. production. In any case, he could do nothing with statistical estimates. He would need hard performance numbers, flight test results. The sleek bird of prey Edgar Schmued had devised, which the American and British military colony in England all fervently supported, fizzled like a wet firecracker before the chief of the Army Air Forces. Arnold dismissed the Mustang. Hitchcock and the two other officers left the Pentagon desolate.

Peggy Hitchcock had flown down from New York to be with her husband in Washington. Back at her sister-in-law's Georgetown town house that evening, Peggy said she had never seen her husband so depressed. Stunned and dismayed, Tommy brooded and hardly slept that night. But when morning came, he reassessed, taking stock of the Mustang's predicament. The fight for the superlative fighter could not fail, he realized. The Allied air war, he felt certain, would soon come to revolve around the astonishing plane. He would not go back to England. Instead, he would take his crusade for the Merlin Mustang around the country. He would stay on in the States until February.

Several weeks later, hard performance figures on the trials of the converted Mustangs arrived from England and California. The plane's extraordinary top speed of 427 miles per hour, its fast rate of climb and other amazing benchmarks confirmed the P-51B could fly like no other plane in the Allied arsenal. On a cold day in December, the numbers landed on the desk of Robert A. Lovett, the undersecretary of war for air and a civilian of enormous importance in the Pentagon. He strode straight down the hall to Hap Arnold's office. The Army Air Forces had to go with the Merlin 61, he told Arnold. They could not waste time with the less powerful Allison engine. "I remember insisting that we could not fool around with an inferior product when a superior one was available," recalled Lovett. "There was no question of what to do, really."

Meanwhile, other forces were at work. On arriving at Wright Field in November, Chesley Peterson had flown all the fighters on hand. But when he asked to fly a Merlin Mustang, the general in charge of testing turned him down. "The P-51 is not in the inventory. It is not in the 'terms of reference.'" This meant that a plane with a British engine, the Wright Field generals agreed, would not be considered. Peterson demanded to see General Arnold: Arnold repeated what he had said to Hitchcock, Lyon and Slessor. "The P-51 is finished," he told him. "The production line has ended. . . . You go back to England." At dinner with Tommy in Washington before his departure, Peterson was disconsolate. "They didn't listen to me," he told Tommy. "They wouldn't let me tell the story." The bodies of Mustang supporters were piling up.

Tommy pressed ahead. Churchill had written to FDR asking him to see Tommy about the P-51. Now Tommy had his private meeting at the White House, and Roosevelt's note to Arnold that same day was to the point: "I am told by a young American friend returning from England that the British are very keen about the P-51 and feel they could use Rolls-Royce engines in them. . . . Do you know anything about it? . . . Can you give me a tip?"

FDR's note reversed Arnold's plans to scrap the Mustang once the British order was fulfilled. Within hours Arnold "sent a memo to General Echols, the general in charge at Wright Field, asking him to fill in numbers of P-51s to be built in the blanks in a rough draft of Arnold's proposed reply to Roosevelt," recounts the aviation historian Paul Ludwig. The figure chosen by Major General Oliver Echols at Materiel was 2,200 Mustangs. The logjam, it appeared, had been broken.

And then nothing happened. There was no production in December nor in January: the factories were quiet. Arnold had given the order, but he had not followed through. As Lovett would say later, "His hands were tied by his mouth. He said our only need was Flying Fortresses." He did not make sure the P-51 got manufactured because he still did not believe in it. In February, again, no Mustangs would be produced. With Arnold doing nothing to advance the project, from November to the following February, "Hitchcock took on the job of seeing that the production order for the P-51 was carried out," his biographer writes. As historian Lynne Olson puts it, he "appointed himself the ramrod of the project."

Tommy did not miss a beat. Like a runner gripping a baton, he took off on an exhausting cross-country shuttle to win acceptance for the Merlin Mustang: to Wright Field in Dayton, Ohio, where Materiel Command was located; to the Inglewood, California, plant where North American was manufacturing the plane; to the Packard factory in Detroit, where the Merlin engine was being produced under license from Rolls-Royce; and back to the Pentagon, crisscrossing the country in planes, trains, automobiles.

Everywhere he went, as he wrote to Peggy from the road, he "spent all day talking airplanes"—talking Mustangs, reasoning, persuading, cajoling, arguing, anything to clear roadblocks and get the production lines rolling. On this one-man pilgrimage, Tommy was not only visiting factories and airfields; he was waging a sophisticated public relations campaign to convince military officials at Wright Field and the Pentagon, the press and the public of the sheer superiority of the plane. His grueling cross-country shuttle was a crusade to win hearts and minds for the Mustang.

One trip was especially challenging. In a letter to Peggy in December 1942 from the Beverly Wiltshire Hotel in Los Angeles, he described how, while traveling in a converted B-24 bomber from Detroit to San Diego, the heat went out, gasoline leaked into the cabin and the radio fell apart in the copilot's hands. When the brakes failed in Tulsa and no repair was available, the pilot took off anyway. On landing in San Diego, Tommy wrote: "There was a little excitement . . . as the left wheel came off." From San Diego he went on to North American Aviation to observe the P-51B on the assembly line. "The Mustang still looks like the best fighting plane we have got," he wrote.

What he saw in his travels was the country ramping up its wartime production. Los Angeles was a city transformed in two years by the aircraft industry, with "thousands . . . who stream[ed] onto the buses and streetcars, and form[ed] endless worms of automobiles . . . going on to the night shift at Lockheed or Vultee," where the "wings of bombers move[d] evenly along an assembly line, and dried hard as rock in a few seconds [under] an overhanging tube of ultra-violet light."

In 1941, as America geared up to make airplanes for its European allies and then for its own war effort, the industry grew 300 percent. Even before Pearl Harbor, overseas orders, as well as domestic orders from a nation that was unprepared for war, led to a kind of modern industrial revolution. Now, in 1942, as America rolled up its sleeves to stoke the war effort, production nearly tripled again. Sleepy Southern California was transmogrified almost overnight as factories such as Northrop, Douglas and Lockheed grew to meet the call of war. Men and women—most of whom had never flown in an airliner—traveled to California from all over the country to join the burgeoning aviation industry and escape the effects of the Great Depression. A huge flood of women who had never before been employed now learned new manufacturing skills. This would have a historic impact on American society long after the war was over.

It was FDR's great Arsenal of Democracy, milling out airplanes like lug wrenches. And yet P-51 production was definitely stalled. Returning to England from a trip to the States, Philip Legarra, North American's representative in London, had reported to Tommy Hitchcock that the Mustang had the lowest priority that could be assigned by the U.S. government to an airplane. Among all the Army Air Forces personnel from Washington to Dayton, few officers seemed to care for the swift, nimble bird of prey Hitchcock knew was so important.

CHAPTER 11

Cash and Corruption

No one, looking back, could identify for sure the source of the USAAF's incomprehensible aloofness to the Mustang, or trace precisely the course of air force intransigence from the autumn day in 1940 when the raptor had first flown. But official reluctance had likely started long before, in the early days when the plane was still the gleaming aluminum NA-73X. In 1940, General Echols at the air force's Materiel Command had suggested to the Anglo-French Purchasing Board, renamed the British Purchasing Commission after the fall of France, that North American would produce Curtiss P-40s for the RAF as a favor to the Curtiss-Wright Corporation. The independent Dutch Kindelberger and his designer Edgar Schmued, with drawings of a revolutionary fighter already in his head, had refused. Some historians claim that North American's reluctance was enough to turn Echols against the Mustang.

Delays in testing the plane had begun almost at once back in 1941. The Materiel Command at Wright Field had received two

P-51As for testing from the first batch of prototypes completed for the British. But month after month, the planes were ignored while other aircraft were given priority. No tests were conducted, no evaluations done, no reports drawn up for a full year and a half after Materiel received its first Mustangs. It seemed the P-51 had collided with a monolith. The outrageous delay would lead one indignant officer, Major General Muir Fairchild at the Army Air Forces headquarters in Washington, to write a scathing letter to Materiel demanding to know the reason for its stonewalling on the Mustang. "The P-51 appears to be the most promising fighter in existence," he wrote. "Its production and increased performance are of paramount importance." Not until July 1942 did the P-51 draw a positive reaction, and then it came not from Wright Field but from Eglin Field in Florida. Materiel had delayed sending the plane to Eglin for routine evaluation. When it finally arrived, Eglin's report gave the Mustang the highest praise for its superior flying characteristics, stability and well-balanced control.

Hitchcock thought the problem was American prejudice against the British. This would not have been surprising. Anti-British sentiment in the U.S. during World War II was in fact widespread. It was certainly not limited to opponents of the Mustang. The American journalist Edward R. Murrow wrote about the common American wartime prejudice against the British, attributing it to "partly the frustration involved by war without early victories . . . and partly the tendency common to all countries at war to blame their allies for doing nothing."

As Hitchcock wrote in a memo that year: "Sired by the English out of an American mother, the Mustang has had no parent in the Army Air Corps . . . to appreciate or push its good points." Tommy observed: "It does not fully satisfy important people . . . who seem more interested in pointing with pride to the development of a

100% national product than they are concerned with the very difficult problem of rapidly developing a fighter plane that will be superior to anything the Germans have." Beyond that objection, since the P-51 had been originally ordered by the RAF, it had not passed through the normal American layers of engineering scrutiny. Essentially, the attitude was "not invented here." In other words, a plane ordered for the British, even if it was designed and built by an American company in the United States, must be inferior. In the words of Harry Hopkins, FDR's top aide, "many Americans believed they naturally flew better than the British and always built better planes than the British." It was pure chauvinism.

Compounding the problem was the inescapable fact that a British engine, the Merlin, was to be substituted for the Allison, an American one. The Army Air Forces through Materiel Command had developed and paid for the Allison; it was now being used in many of the American planes being sent into battle. In addition, "the Allison-powered Mustang had been extensively publicized in the press, somewhat misleadingly, as an American plane, a brilliant contribution of American technological genius to the war effort. The problem was to explain to the public that the Mustang was indeed a splendid aircraft, but that it could be made just a little better with a Rolls-Royce engine," the historian Paul Kennedy writes.

Tommy, ever the promoter, used the press to make his case to the American public for Britain's participation in the P-51. In one interview he was quoted as saying diplomatically: "We are drawing freely and indiscriminately on the engineering skill and productive power of both England and the United States to best the Germans in the race for air superiority."

Much of the problem could likely be traced to Wright Field, the sprawling testing and development complex of the Army Air Forces, spread out like a giant star field near Dayton, Ohio, where "made in America" was more than a tag; it was a burning faith.

Wright Field had its origins in the earliest stirrings of aviation. After their first successful flight at Kitty Hawk, North Carolina, in December 1903, Wilbur and Orville Wright returned to Dayton and bought an 84-acre pasture to use as an experimental flying field. At Huffman Prairie, Orville wrote, they really learned to fly, perfecting their Flyers II and III. In 1910 they established the Wright Company School of Aviation there—among its pupils, of course, was Hap Arnold. When America entered the First World War in 1917, the best airplanes were European, and the U.S. went to war in Bréguets, Nieuports, Salmsons and de Havillands. But in 1917, the U.S. Army purchased the Wright brothers' pasture and the 2,000 acres next to it, and Wright Field soon became known as the leading innovator in aircraft development, from engines to airfield improvements, in the shift from biplane to monoplane wings.

During World War II, Wright Field grew from 30 buildings to some 300 and boasted the first modern paved runways in the Air Corps, by then called the U.S. Army Air Forces. It bulged from under 4,000 employees as 1940 began to over 50,000 at the war's peak.

Most important there during World War II was the air force purchase and procurement arm—the Materiel Command. It would be responsible for buying new aircraft and aerial equipment in production quantities and for a stepped-up program of testing and development. For the most part, Materiel aptly handled the enormous job of choosing, developing and dispatching dozens of different aircraft to the front. Historians praise its achievement in the fiery trial of a global war: it created a modern air force almost from scratch, growing exponentially in size and going from developing and supervising manufacturing on less than 750 airplanes in World War I, to fathering a force totaling 41,000 planes by the end of World War II, including scores of widely differing airplane types. The Materiel Command ordered the prototypes that won the twin sky wars in Asia and Europe. But in the process, decisions made at the top were often influenced by prejudice and by vested interests.

At the helm of Materiel, the stiff-necked Major General Echols presided over the ordering and proving of all planes for the U.S. Army Air Forces. He was the central figure in the procurement of aircraft throughout the war, when American industry was called on to produce an unheard-of number of aircraft: an overall total of 300,000, including aircraft of all types made in the war. Aircraft production tripled twice under Echols, with 50,000 planes a year being built at war's end: he was a vital engineer of the Allied triumph in the air in both Asia and Europe. But on the Mustang, Echols erred badly. His relentless anti-Britishness and that of others at Materiel was well known. Tommy's perception was correct: the facts that the P-51 was made for the British and that it would now be powered by a British engine were two strikes against it.

But beyond this a system of cronyism ruled at Wright Field, putting favored airplanes with strong backers at the head of the line for contracts. A sign of the clout established manufacturers could wield: in the fall of 1942, during the Mustang's exile, with the American war in Europe well under way, Materiel had on order for the Army Air Forces 2,500 Curtiss P-40 Kittyhawks, no fewer than 8,800 Bell P-39 Airacobras and a boggling 11,000 Bell P-63 Kingcobras. None of these aircraft was any match for the German Focke-Wulf Fw 190s then coming online. They were all outdated. In addition, over 5,000 Glenn Martin B-26 Marauders, called "widow makers" because they crashed so frequently during takeoff, continued to be made despite their defective wings. Curtiss-Wright in particular was heavily favored by the air force; it was automatically handed new contracts while other companies pleaded for business.

General Michael P. C. Carns, a retired four-star general who served as vice chief of staff in the air force, observes that this behavior is not uncommon in the defense contracting arena. "It's going on today. These people are highly reliant for their futures on getting continued contracts. And so there are major efforts put forward by all the germane manufacturers," says General Carns. He tells a story that he cannot confirm but that illustrates the politics that can color defense procurement.

"When the F-111 [attack aircraft] contract was being bandied about, there were strong rumors that the Defense Department decision had been made to recommend Company X to the White House as the winner of this major competition," Carns recounts. "At the time, Lyndon Johnson was the president. The appropriate briefings and recommendations were submitted. The White House announced Company Y, a Texas enterprise, to be the winner, a major surprise," he relates. "The street dialogue was strong; rumors abounded. Everybody said, 'That's not the plane we selected.' The new winner was General Dynamics, from Texas." As was often the case, politics determined the selection.

But this time the generals were playing favorites with an airplane that could turn the course of the air war. They were ignoring the tests in England and California that showed the Mustang to be the world beater it was. They were throwing under the bus a plane unprecedented in fighter design, with capacities of speed, range and power that could bring victory to the Allies.

At the root of the favoritism at Materiel, the source of $60 billion in contracts through the war, there may have been something more sinister than the kind of influence peddling General Carns describes: the nefarious dealings of one Major General Bennett E. Meyers. "Benny" Meyers, a "peppery little man with a big cigar in his mouth and a brigadier's star on his shoulder," was the second-in-command at Materiel from 1940 to 1942 as the Mustang was evolving from the P-51A to the Merlin-engined P-51B, and he held numerous positions at Materiel throughout the war. He also had a little business on the side. It was a metal tooling factory that made electrical parts for aircraft, a small operation a few miles from Meyers's office at Wright Field in Dayton—small but very profitable. Meyers's company, Aviation Electric, with just five employees running thirty machines, sold minor items such as fuse boxes to a number of companies with which the Army Air Forces did business. The companies that paid his exorbitant prices throughout the war, a total of $20 million in today's dollars, received contracts for their aircraft from Meyers.

Supervision at Materiel was light: Meyers's boss, General Echols, delegated many contracts to him without oversight. And Meyers took great pains to cover his tracks. He made his strikingly beautiful brunette secretary and her mild-mannered husband, Mrs. and Mr. Bleriot Lamarre, the titular secretary and president of Aviation Electric. On paper the company did not belong to him. But behind the scenes, Meyers was pocketing Aviation Electric's huge profits— and, in a further boondoggle, was buying stock in the aircraft companies to which he awarded contracts. The money funded a lavish lifestyle, including a $3,000 Cadillac, an $825 air conditioner, a $700 radio, and a $10,000 decorating job on Meyers's Washington apartment, as well as contributions to the Lamarres' lifestyle, as Mrs. Lamarre was his mistress "with Mr. Lamarre's knowledge, approval and acquiescence." All in all, it was "a cozy setup for the cozy war Meyers was fighting in Dayton and Washington," commented *Time* magazine.

Lawrence Bell of Bell Aircraft became an early client after Meyers told him he could recommend a company to make electrical airplane parts that was "owned by some friends." Eventually Aviation Electric did more than a million dollars of business with Bell between 1940 and 1945. Other aircraft companies soon bought parts and received contracts, too: Republic ($326,587), Curtiss-Wright ($10,002), Glenn Martin ($19,263), as well as the Vimalert company, which made parts for the Allison engine. North American was notably absent from the list of those companies that bought parts from Aviation Electric and got contracts from Meyers. More than one historian has suggested that Dutch Kindelberger's failure to play Meyers's game cost him a major contract for the P-51 until late 1942.

Wright Field was not entirely oblivious to Meyers's dealings. In the spring of 1940, General Echols, hearing gossip on the base about Meyers's enterprise, asked him to sever his ties to any company that was doing business with the government. But Echols did not follow through. Again in 1943, the U.S. Army Air Forces high command was concerned by rumors that certain obsolete or under-performing airplanes were being favored by officers at Materiel because those officers owned stock in the aircraft companies that made them. These included the Curtiss P-40 Warhawks, Bell P-39 Airacobras and P-63 Kingcobras and Glenn Martin B-26 Marauders, the "widow makers." The chief of the Air Staff, Lieutenant General George C. Stratemeyer, sent out a questionnaire to high-ranking air officers asking for a statement of their aircraft security holdings. But when Meyers reported that he owned no aircraft stock—literally true, as he had already transferred his investments to his wife's name—the investigation was dropped.

Meyers's next shady venture was in bonds. By the end of the war, he was speculating on a large scale in government war bonds in violation of explicit government policy. This illegal activity was what finally brought him down. In the spring of 1944, Howard Hughes, the famed pilot and motion picture producer, who also manufactured airplanes for the government during the war, was hoping to sign a contract with Meyers to develop a plane called the F-11. Meyers insisted on an advance kickback on the contract in the form of a $200,000 personal loan so he could buy war bonds. When Hughes refused him, Meyers found the money elsewhere and proceeded to buy more than $3 million worth of war bonds in 1945.

Fast-forward to 1947, when Hughes was being investigated for the same $40 million F-11 contract—which in the end had yielded not a single airplane for the government. As part of that investigation by the Committee to Investigate the National Defense Program, better known as the Truman Committee, the head of the committee, Senator Homer Ferguson of Michigan, ordered a close check into Meyers's wartime activities.

The committee got its start in 1940 when Senator Harry Truman set out from Washington on a cross-country trip in his old Dodge automobile. He had heard about waste and corruption in the defense industry preparing for war and was determined to see for himself. Driving more than 10,000 miles from Florida to the Midwest, he visited military plants and saw firsthand widespread abuse of government dollars by a handful of military contractors. "The experience," he said, "was an eye opener." His biographer David McCullough writes: "Slightly built, bespectacled, a lover of Chopin and a shunner of the limelight," Truman struck most people as anything but a crusader. Yet he got results. In March 1942, Congress passed Resolution 71, unanimously giving Truman a modest $15,000 to set up his committee on defense contracts.

Committee members "travelled the length of the country, putting in at a city or military base. . . . War plants were inspected, hearings held in local hotels." Holding more than 400 public hearings from 1941 until its end in 1948, the committee found and corrected problems of waste, inefficiency and profiteering throughout the defense industry. They proved wartime shortages, cost overruns, nepotism in hiring and labor strikes. Perhaps most notorious was the case of the Curtiss-Wright Lockland plant, which made engines for B-29 bombers. The committee found that a corrupt inspection process where inspectors were discouraged from rejecting defective engines resulted in bombers crashing and killing their crews. This case was memorialized in Arthur Miller's Broadway play *All My Sons.*

In a search of government files in connection with the Hughes scandal by the Truman Committee, an anonymous 1945 letter was found addressed to the FBI; it asserted that Meyers had made huge profits by buying stock in aircraft companies to which he was about to give air force contracts. Specifically, the letter said it was Meyers's "practice as soon as he was in possession of the bids for review to buy stock in the particular company prior to the release of the bid."

In looking into Meyers's income tax returns and his bank statements, Truman Committee investigators discovered the checks to Aviation Electric from aircraft companies and records of Meyers's stock holdings in those companies. As 25 wounded GIs, many with empty sleeves or trouser legs and wearing their medals from the war, listened intently in a congressional hearing room, General Arnold, who had "stomped into the hearing after a hurried flight from his ranch in Sonoma, California," according to one journalist, said Meyers had "disgraced his uniform and his rank," but called him only one "rotten apple" in the air force barrel. Meyers was soon convicted on three counts of suborning perjury. The air force stripped him of his medals, rank and retirement pay and prepared to haul him before a court-martial.

Meyers was in fact a superbly efficient general officer. Through the war he held various top positions at Materiel, handling the actual purchasing of air force supplies and equipment. During the war years he oversaw aviation contracts involving billions of dollars. The air force brass saw him as critical to running the war and depended on him to cut through red tape and make things happen. Arnold described him as the "spark plug" of air force procurement. Very simply, Arnold said, "he got things done."

Busy building an air force and then running an air war on both sides of the globe, and depending on Meyers to help him succeed, Arnold was likely not a party to his schemes or even aware of them during the war. In any event, he was not confronted with Meyers's wrongdoing until 1945 when the war was over.

But if Meyers's preference for aircraft companies who bought into his schemes was causing the air force to prop up those companies by awarding them contracts for inferior planes in preference to a fighter that could contend with anything the Germans put in the air; if, as some historians suspect, North American's refusal to patronize Aviation Electric cost Dutch Kindelberger a contract for the Mustang long beyond the time the plane's merits were well known—then something far darker was taking place. If these allegations are true, one of the most critical weapons of World War II was being stymied by vice and corruption.

But there was another reason for Hap Arnold's lack of interest in the P-51. Conventional wisdom at the time among officers of the USAAF dismissed the importance of fighters. The reason stemmed from an old and outdated law of aerial warfare called "bomber theory."

Bombers, the theory went, would dominate all air war. Such had been the thinking ever since the beginnings of aerial bombardment, in the Italo-Turkish War of 1911, when an Italian pilot dropped a hand grenade from his plane. For decades, "bomber theory" ruled the thinking of air force generals, shaping every decision and strategy. Bombers were the principal means of attack: the destructive power of bombers was the centerpiece of all air war. Flying fast and high, the bombers—bristling with tail guns, waist guns, nose guns and belly guns, and drawn together in a tight checkerboard grid—would form an interlocking, invincible fortress in the sky. Bombers would always reach the target. There was no need for fighter escort. Fighters were superfluous.

In any case, the argument went, a long-range, high-performance fighter could not be built. It was an impossible dream. Conventional wisdom ordained that if one developed a long-range pursuit aircraft with extra fuel tanks to extend range, the weight of the tanks and additional fuel would slow the plane and critically hamper its performance.

Or so generations of officers from the early days of aerial combat up to World War II had thought.

They were disciples of the legendary Billy Mitchell, the acknowledged founding father of American airpower. Learning to fly only at 36, Mitchell was appointed chief of the brand-new Aviation Section of the U.S. Army Signal Corps. The first American airman to arrive at the Western Front in World War I and the first to fly over enemy lines, he was idolized by many young American pilots.

Drawing his ideas from aviation visionaries like British air marshal Hugh Trenchard, first commander of the Royal Air Force, and the Italian luminary General Giulio Douhet, whose treatise on strategic bombing, *The Command of the Air*, captured the popular imagination in the 1920s, he developed the idea that the slaughter on the ground of debacles like the First World War could be forever ended with airpower by using air strikes on industrial targets by massive bomber forces. Airpower could end, or at least shorten and make more humane, all wars. Mitchell became a crusader for the concept of "strategic bombing," antagonizing many. He was later discredited and court-martialed by his opponents in the air force—but by then younger officers had taken up his crusade. Among his disciples, known as the Bomber Mafia, were Hap Arnold, Tooey Spaatz and Ira Eaker.

Hap Arnold was a true believer. His opposite number in the Royal Air Force, Arthur Harris, also worshipped at the shrine of the bomber and its importance in aerial warfare. Both were ferocious exponents of B-17s, B-24s, Avro Lancasters and Bristol Blenheims.

The Boeing B-17 Flying Fortress was the linchpin of American bomber theory. A stronghold in the sky, it looked like a huge fish, though majestic to the eye. It had grown out of an early Boeing project in 1935 called the Model 299 and its ultimate form, morphing through the B-17C, D, E and F, became the B-17G. The G was the hammer of the USAAF. It was capable of speeds up to 300 miles per hour on four 1,200-horsepower Wright R-1820-97 Cyclone engines. Studded with no fewer than eight gun positions, mounting twelve Browning .50-caliber machine guns on nose, tail, waist and belly, the G had an immense bombload capacity of 17,600 pounds, about nine tons. It had a combat radius of up to 800 miles. Mated with the Norden Mk-15 bombsight, it made a fearsome weapon.

The Fort was teamed with the Consolidated B-24 Liberator, its partner in the Eighth Air Force, but with a lighter bomb payload, 8,800 pounds, and a shorter range. The Fort and the Liberator were both tough, mighty sky rams, but the fortunes of war would show them to have a serious vulnerability. Their own guns, in planes holding steady and level, were insufficient against heavy fighter defenses. While they could fly high and fast, their need to maintain a straight, unwavering course to drop their bombs and their lesser maneuverability made them easy prey for Nazi attackers. Both would depend on a fighter escort like the Mustang; without them they would fail.

And the present lineup of Allied fighter escort was wholly inadequate—unable to fly the vast distances of bombing raids—to provide this guard. The Army Air Forces desperately needed a long-range, high-performance plane.

But with the forces of anti-British prejudice, vested interests and possible rank corruption arrayed against it at Wright Field and compounded by the misguided notion of bomber theory, the Mustang sat on the sidelines, cast aside through 1942. Official U.S. Air Force historians have observed that "the story of the P-51 came close to representing the costliest mistake made by the AAF in World War II."

In 1943, however, an event took place whose repercussions would blow the shibboleth of bomber theory wide open, turn the air war on its head and finally bury the myth of bomber invincibility. It would usher in a terrible passage of time when the air war would hang suspended and bomber crews would be lost by the thousands, and finally lead to a headlong rush to produce a fighter plane that could accompany the bombers all the way to their targets. The drama that year began not amid the black puffs of flak in the skies over Bremen and Muenster but in the shade of palm trees and exotic gardens as Allied leaders met to plan the endgame of the war.

CHAPTER 12

Huddle in the Orange Groves

On a cold black night in January 1943, a motorcade of cars with flashing headlights and blaring sirens bore Winston Churchill and his entourage to a secluded RAF base near Oxford, England. When his car arrived, Churchill, in the uniform of an air commodore, lingered on the runway as the plane was refueled, finishing his cigar and waiting amid the tight security. Then he turned and boarded his personal B-24 Liberator, *Commando*, converted from bomber to government transport. With him were top commanders from his military staff, their destination a secret.

Gunning the Liberator's massive Pratt & Whitney radial engines, the pilot peeled away down the blacked-out runway and lifted off into the damp English night, climbed to 7,000 feet and then swung south for a nine-hour flight that would take Churchill and his generals to the city of Casablanca in Morocco. There he would meet President Roosevelt, who would arrive after a five-day journey by Boeing flying boat and Army C-54 transport. Their meeting would be code-named "Symbol." Its purpose: to plan the last stages of war.

That January, Tommy Hitchcock was crisscrossing the United States, promoting the P-51B Merlin Mustang; Nazi U-boats were sinking hundreds of thousands of tons of shipping on the bitter North Atlantic; the land war was stalled in Russia and the Mediterranean. On this meeting in Casablanca would depend the fate of the free world, hanging now on the struggle between the Axis and the Allies.

Inside the plane, at 30,000 feet, the air was frigid. Most of the passengers had bedded down on makeshift bunks, wrapped in layers of heavy clothing. The exception was Churchill, who slept dressed in a silk nightshirt. A kerosene heating system had been installed in *Commando*, but Churchill and Sir Charles Portal, the top commander of the RAF, thought the system dangerous and likely to start a fire. So they disabled it and bedded down again, chilled and sleepless. The B-24 Liberator droned south over the dark Atlantic in the isolation of the vast winter night.

Early the next morning, the plane dropped out of the high clouds and swooped in over the native felucca boats swinging at anchor and the cadmium white yachts stranded by war in the harbor. In the distance lay the vista of the snowbound Atlas Mountains to the south; below, the flowers, palm trees and orange groves, the minarets and whitewashed houses marked the port city of Casablanca.

To secure Symbol, then-Brigadier General Dwight D. Eisenhower had commandeered a site in the coastal enclave of Anfa on Casablanca's outskirts, taking over 18 private villas and the unpretentious but very modern Anfa Hotel for lodging and conference rooms. The Americans had surrounded the site with barbed wire fences enclosing the compound—about a mile in circumference—and set a guard of hundreds of American soldiers armed with machine guns and bayonets, standing on rooftops and patrolling walkways. Blackout curtains were pulled across the hotel windows "because Casablanca was well within range of any Nazi bomber that might be based in Spain or southern France," wrote General Eaker's aide James Parton, who attended the meeting. Here, near the colorful medinas, mellahs and riads of Casablanca's ancient neighborhoods, the conference would unfold.

FDR arrived two days later. It was his first trip in an airplane since he had flown to Chicago in 1932 to accept the Democratic nomination for president and his first by air over the open ocean, a circuitous journey with stops at Miami, Trinidad, Brazil and then, after eighteen hours in the air, at Bathurst on the Gambia River in West Africa before the final leg to Casablanca. No president had ever before traveled by plane, least of all one in Roosevelt's poor health. His doctor carried digitalis on board in case the altitude should trigger a heart attack.

Symbol lasted from January 12 to January 23. For the top secret meeting, FDR went by the code name "Admiral Q," Churchill by "Air Commodore Frankland." They were quartered in neighboring villas 2 and 3—FDR's villa with a bed "at least three yards wide and a sunken bathtub in black marble," according to Churchill historian Andrew Roberts; their staffs were lodged in the Anfa Hotel. On hand was "almost every conceivable personality in the armed forces of the American people and the British Crown of whom one has ever heard . . . along with great hosts of marines, WRENs, WAAFs, typists, telegraph officers, etc. of all kinds," wrote Harold Macmillan, then a young Churchill aide and later himself prime minister. In the bay of Casablanca stood a command communications ship capable of sending 30 wireless messages at the same time, staffed with a flock of coding experts.

Here, with their phalanxes of support personnel, adjacent to the balmy, colorful sprawl of Morocco's principal city, the two Western chieftains met to survey the map of the war and chart how best to deploy American and British troops to defeat the Axis enemy— spread out like an evil shroud around the globe. For the next eleven days, the Allied staffs, headed by the American and British combined chiefs of staff, met by day in a long conference room in the Anfa Hotel with windows on one side looking out on palm trees and flowers, a pleasant change from blacked-out London and from wintry Washington. When the workday ended, "you would see field marshals and admirals going down to the beach to play with pebbles and make sand castles," recalled Macmillan. They walked by the shore and swam in the sea; Churchill particularly enjoyed the sand and surf of Anfa. Macmillan would write in his memoir that the atmosphere was "a curious mixture of holiday and business in these extraordinarily oriental and fascinating surroundings."

The generals and flag officers then briefed the leaders at FDR's villa from evening until late into the night, "the scene [often] lit by candles after an air raid alert forced the dousing of the electric lights." Churchill kept to his usual routine of spending much of the day in bed, drinking heavily and then dining with Roosevelt before the nocturnal briefings began. When Roosevelt retired he would turn, as was his habit at home, to furious bouts of paperwork far into the night.

The Casablanca conference laid out the master plan for Allied military operations from 1943 until the end of the war. It reaffirmed the need to give available aid to Russia; to take Sicily in an invasion of Italy, the "soft underbelly of Europe"; and in the Pacific to retake southern Burma to open up a supply route into China. Most important, Roosevelt and Churchill agreed on a "Germany First" policy—the destruction of Germany as the first priority of the war. While the American generals had arrived at the meeting focused on defeating Japan, after tense early days of heated discussion with the British, they left reconciled to defeating Germany first.

Here the wings of aviation would be critical. Winning the battle against the Nazi U-boats on the Atlantic was a top priority, and at sea, airplanes were essential in depth charging and destroying German submarines that lay in wait for the Allied supply convoys. By the end of the war, 70 percent of U-boat kills would be scored by aircraft, principally B-24 Liberators like Churchill's converted *Commando*. And the Allies must have command of the sky to undertake a cross-channel invasion: Allied bombing of German industry and the fight to destroy the Luftwaffe would be the means to that end. D-Day would not be feasible until it could be accomplished.

And so one of the key blueprints laid out at the conference was the Casablanca directive and the Combined Bomber Offensive that emerged from it. The Americans preferred precision daylight bombing; the British held out for bombing at night. Churchill seemed immovable. Hearing a rumor that Churchill had talked FDR into joining the RAF in night bombing, Arnold urgently summoned General Eaker from England to make the American case. Eaker recalled: "Mr. Churchill motioned me to a seat on the couch beside him and began to read, half aloud, my summary of the reasons why our daylight bombing should continue. . . . At one point, when he came to the line about the advantages of round-the-clock bombing, he rolled the words off his tongue as if they were tasty morsels." Handing the memo back to Eaker, he said: "How fortuitous it would be if we could, as you say, bomb the devils around the clock."

The barrage was on; the war now centered on the bombers. Germany would be assaulted 24 hours a day and, weather permitting, seven days a week, the Americans in daylight, the British at night. This offensive, code-named "Operation Pointblank," would target the Nazi aircraft industry, the means to make the Luftwaffe's planes and the fuel to fly them: aircraft parts and ball bearing factories, rubber and tire factories, oil production and storage plants. The new campaign would take RAF and USAAF bomber crews deep into the German heartland.

After the conference broke up, the villas emptied, the ranks and rates flew home and the soldiers pulled out, the summit ended on a grace note. On January 23, at Churchill's insistence, he and Roosevelt drove five hours south to Marrakech. Churchill loved the city, the "Paris of the Sahara," from earlier trips there in the thirties, and now he was determined to show the American president "the most lovely spot in the world." Their destination was the Villa Taylor, an elegant casbah with grounds planted with lush orange and olive groves, small trees and flowering shrubs, built by the American millionaire Moses Taylor and his wife, Edith Bishop Taylor, in the mid-1920s. The house was now the home of the U.S. vice consul Kenneth Pendar. Its gardens of spider ferns, jasmine and gardenias and finished with fountains were dominated by a three-story tower.

On arrival Churchill insisted that FDR, crippled by polio and confined to a wheelchair, be carried up the three winding flights of stairs to the roof of the tower to see the breathtaking view of the land stretching away to the Atlas Mountains. Two servants locked arms, and with Roosevelt's arms around their shoulders, they carried him to the top of the turret. There the two leaders watched the horizon as the sun sank low, sloping down from the snowy Atlas Mountains to the Atlantic, turning the sky from crimson to apricot to fading blue.

The two finished the evening with a lavish dinner of lobster and filet mignon. Later Churchill painted a picture of the view from the tower, the Koutoubia Mosque in Marrakech with the Atlas Mountains behind it, his only picture from the war period, and gave it to Roosevelt. The next morning, Churchill went out to the Marrakech airport to see Roosevelt off, "dressed in velvet slippers and his green-and-red-and-gold dragon dressing gown." At dusk he would depart in his *Commando* for Cairo, the plans for the Combined Bomber Offensive and the last phase of the war set now by the chieftains and their generals.

As Symbol ended, Tommy Hitchcock was preparing for his return to London. Throughout his travels around America to sell the Mustang, he had sandwiched in brief visits with his family in New York. They had spent Christmas 1942 together at the Gracie Square apartment, and he could see that their five children—stepson Alex, teenage daughters Louise and Peggy and four-year-old twin boys Tommy and Billy—were growing up quickly without him. From London he would soon write to his wife: "Seeing you and the children again [has made] separation all the harder. . . . No war, no flying, no Mustangs can replace you."

Certainly he left America with some feeling of satisfaction. Thanks to his cross-country crusade, the Merlin Mustang was finally beginning to come off the assembly line at North American, albeit at a crawl. Production figures told the tale: by March, 70 Mustangs would leave Inglewood; by April and May, 121 each month. The numbers were not yet large, but it was at least a start. Perhaps the massive logjam of the air force bureaucracy had been broken.

But Hitchcock must also have felt some foreboding. He must have wondered if the new Mustangs could arrive in time to make a difference. At a War Department briefing before he left, Tommy had warned that the air war was changing. American raids over France through the fall of 1942 had involved losses that were bearable—

for now. But he knew the Germans were beefing up their fighter forces and that as soon as the Combined Bomber Offensive took the Flying Fortresses into the heart of Germany in the spring, the Luftwaffe would send its fighters in large numbers against the Allied bombers.

Sure enough, there was both sunrise and midnight ahead for the Allies after the Casablanca summit. In February 1943, the Russians would win the Battle of Stalingrad and begin pushing back the Germans to turn the tide on the Eastern Front. And by May, U-boat losses would skyrocket suddenly, leading the Germans to conclude they had lost the war on the Atlantic. But Operation Pointblank, the Combined Bomber Offensive, would hurtle off that summer and soon veer into debacle—and then disaster. It would go more wrong than Allied leaders could ever have imagined, and by late summer scores of American bombers would be shot down like so many coveys of quail. Casualty rates would soar to unacceptable levels, with terrible losses of both airplanes and young crews. The bomber war would become a spectacle; the myth of bomber invincibility would be shattered. By the fall of 1943, the U.S. Eighth Air Force would be on the ropes. The Nazis and their Luftwaffe would command the skies over Europe.

CHAPTER 13

Picked Off like Geese

They flew straight and four miles up, each plane's four 1,200-horsepower Pratt & Whitney radial engines thrumming like kettledrums, droning on. One hundred seventy-eight B-24 Liberators held locked in formation, flying fast, each bomber defended by ten .50-caliber machine guns. Most important, each plane hauled 4.5 tons of bombs. Coming in off the Adriatic Sea over the coast of Greece, their target was the serpentine oil-refining complex at Ploeşti, Romania, which produced 60 percent of Germany's crude oil. To obliterate it would, in one blow, cripple the Third Reich.

Two groups from the Ninth Air Force, the 98th and the 376th, and three more from the Eighth, the 44th, the 93rd and the 389th, soared over Greece and the remnants of Nazi-occupied Yugoslavia. Their mission was planned as a low-altitude operation, close to the ground at treetop level, to avoid radar detection and reduce the target profile of the attacking planes. The raid depended on surprise, so the bombers homed in under complete radio silence; they had no idea the Germans had a radar receiving station in Greece that had already picked them up. Unknown to the Americans, surprise was already blown. The German citadel at Ploeşti knew every move the bombers were making.

It was August 1, 1943.

It had been a high-endurance, long-range 1,400-mile trip to Romania, a seemingly endless seven-hour flight from the American base in Benghazi, Libya. Now the bombers dropped to just a few hundred feet over land. Some of them made a wrong turn into a valley, scattering hopes for a disciplined, coordinated strike. Worst of all, no intelligence flights had been made over Ploeşti; none of the American raiders knew about the defenses at the sprawling refineries. Colonel Alfred Gerstenberg, the Luftwaffe commander, had reinforced the complex like a medieval castle redoubt, with more flak guns than protected Berlin, machine guns, fast-firing cannons and a force of 250 fighter planes armed and ready. The 389th Bomb Group, the "Sky Scorpions"; the 93rd, "Ted's Traveling Circus"; and others were flying directly into a loaded, tightly sprung trap. And they had no fighter escort to defend them; no fighter had enough range for the long journey.

The bombers did not fall aside or turn but flew straight on to their assigned aiming point, the Steaua refinery, coming in low now, just above the trees, steady, straight and smooth. "We were very close behind the second flight of three ships," recalled Captain Philip Ardery, a pilot in the 389th Sky Scorpions, in his memoir. "As their bombs were dropping, we were on our run in." Ardery's squadron droned on through the confusion, smoke and fire, nosing in. They flew smack into a hail of gunfire, mostly 20-mm automatic weapons. The first planes homed in, loosed their bombs. A series of explosions rocked the air: these were the refining facilities blowing up and dumps of volatile gases. Then Ploeşti became a pyrotechnic show. A big central boiler was hit, detonating sky-high in an instant. Flames shot up in the air. The next trio of B-24s went in and hit the boiler house. More explosives, with towering sheets of flame. A mass of black smoke curled up, darkening the cindered sky.

Ardery's radio operator called to him that the aircraft on his right, their wingman, had been hit and was leaking fuel on his left wing. This was Lieutenant Pete Hughes, a religious man who had a young wife back in Texas. Ardery looked and saw the gasoline. He knew Hughes was flying into a solid wall of flames with gas trailing from his ship. He was finished. Ardery said a quick prayer.

Then bombs away.

Long sticks of bombs dropped from Ardery's bomb bays. The sky was blacked out for one eternal moment and they flew through darkness. They cleared some chimneys by a few inches. An explosion kicked up the tail and forced down the nose of the Liberator. They emerged into daylight. Now Ardery could see Hughes next to him: the Liberator turned up and out, having expended its bombs, trying to slow speed for an emergency landing. Then, just before landing, the left wing of Hughes's stricken B-24 was suddenly torn off and the heavy ship cartwheeled groundward. It spiraled, fell out of the sky and crashed into the Romanian loam. Hughes and his crew were gone. They had not hesitated, had never backed away or peeled off, though they knew they were mortally stricken. They had carried their bombs to target and struck hard. Mission accomplished. Objective achieved.

Now the sky was mass havoc, bombers flying in many directions, some on fire, some with smoking engines, some with huge gaping holes in them. Later Ardery would write: "Many [planes] were so riddled it was obvious their insides must have presented starkly tragic pictures of dead and dying, of men grievously wounded who would bleed to death before they could be brought any aid; pilots facing the horrible decision about what to do—whether to make a quick sacrifice of the unhurt in order to save the life of a dying man, or to fly a ship home and let some crew member pay with his life for the freedom of the rest."

Ardery and his crew flew on through the skyway of smoke, cinders and licking flames. Then out of the target zone, out past Ploești, up above the town of Câmpina, off over the receding fields and farmlands, past a railroad station in the midst of the sprawling countryside, back out of Romania and finally over Greece and the coast and the unbroken emptiness of the vast, still Adriatic Sea.

Just nine of the 389th's bombers struggled back to base and landed in Benghazi that night as the sun set. Of the 178 bombers that had departed on the raid, only 33 survived or were fit to fly the next day, almost an 80 percent loss rate. Ploești had been a debacle. Three hundred ten men were killed or missing, many more wounded. The raid had been a disaster.

The Nazis corralled 10,000 slave laborers and repaired the Ploești petroleum complex within weeks. The Allies did not bomb the Romanian facility again until April 1944, and then from high altitude. The Ploești raid was the culmination of a summer of loss for the Eighth Air Force. It would go down in Army legend as "Black Sunday." As the Combined Bomber Offensive picked up steam, Tommy Hitchcock's grim vision of the future was becoming reality: without a fighter shield, the bombers were hopeless victims.

Hitchcock had known all along that a high-performance pursuit plane with a longer range was vital for the bombers. But no such fighter was in use in 1943. As the Americans had begun to operate from England in 1942, the rugged Republic P-47 Thunderbolt arrived. But the P-47s, called "Jugs" because they resembled milk bottles, were unwieldy and climbed poorly. Worse, their range was too short—from their American bases in England the P-47s could

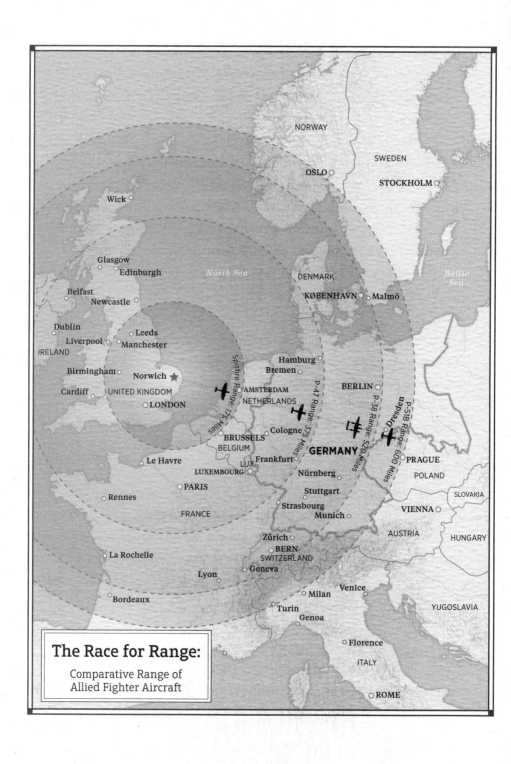

The Race for Range:

Comparative Range of
Allied Fighter Aircraft

fly only as far as the German border. Then they were forced to turn back, leaving the bombers exposed in the treacherous skies over mainland Europe. Chaos would follow.

In the summer of 1943, the Lockheed P-38 Lightning appeared. The Lightning was a more agile, gymnastic plane. A stellar performer in the Pacific, it was the favorite of Major Richard Bong, the top American ace, who racked up a grand total of 40 kills in it against the Japanese. And its longer range—twice that of the P-47—allowed it to stay with the bombers into Germany and back. But the Lightning had an Achilles' heel, a fatal malfunction. Frequent engine failures made it a poor bet in the chilly European skies. In the icy air of the ceiling at 30,000 feet over northern Europe the engines would detonate, damaging pistons, rods and crankshafts. In other cases, engine parts would freeze, lowering the octane of the fuel and resulting in fouled spark plugs. In the end, more P-38s were being lost to engine problems than to enemy aircraft. Their unreliable engines were forcing many Lightnings to turn back far short of their targets.

As Operation Pointblank went on, the B-17 Flying Fortresses would leave the English coast accompanied by Thunderbolts or Lightnings, but when engine trouble forced the P-38s to withdraw or the P-47s reached the limit of their fuel, the escorts turned back. Neither aircraft had the internal fuel capacity to fly deep within Germany. The use of auxiliary fuel tanks to increase a fighter's range began at this point in the war. These were "drop tanks," external wing tanks or belly tanks that could be jettisoned after their fuel was consumed. Faced with wartime metal shortages, the British invented drop tanks made of papier-mâché: heavy paper impregnated with glue. Though the U.S. used more than 15,000 of the British tanks, American drop tanks were made of aluminum. One-hundred-eight-gallon to 150-gallon drop tanks mounted under the wings, and shucked off when empty, offered great hope for extending the range of all the fighters and were used by the P-51 Mustangs. But they were insufficient to adequately increase the reach of the P-47s and P-38s, whose capacity was far smaller.

As the Flying Forts and Liberators pushed on alone, unescorted, into German airspace, winging in over the midlands and the industrial heartland of Germany, the game was always the same. The Luftwaffe pounced. As the bombers roared on to their aiming points, closing with their bays full of 500-pound bombs, they were picked off like so many flights of geese in hunting season, with thousands of young airmen going to their deaths.

Over Bremen in June 1943, bombers reaching their targets would be badly mauled. In August, weeks after the Ploești raid, over ball bearing factories in Schweinfurt and Messerschmitt assembly plants in Regensburg, again nearly 19 percent of Allied planes would be destroyed. Of those bombers that returned to base, roughly half would be severely damaged. Without fighter protection over the target, the Americans would lose something like 75 percent of their aircraft. The cost in human lives would be even more terrible. Hardly a plane would come back without a cargo of dead and mutilated men. Such losses were unsustainable.

"Bomber bases were damn depressing places. Death was always in the air, even though the guys were trying hard to laugh and forget," wrote the journalist Andy Rooney.

But it was not just the massive physical toll the raids took in lives lost. There was also a huge psychological cost. In the air war over Europe in 1943, it became clear to Army Air Forces doctors that pilots and aircrewmen were fighting an entirely new kind of war, with stresses never before encountered since the day of spear and arrow. This new kind of battle in the frigid atmospheres of the firmament took its toll in several different ways. Early Army Air Forces doctors, under Colonel Malcolm G. Grow and Colonel Harry G. Armstrong of the air force's Central Medical Establishment (CME), determined they were going to need to treat the psychological injuries of fliers—the symptoms of strain and emotional exhaustion suddenly appearing in Europe—as well as their physical wounds.

If air war was new, so was its particular brand of harshness. The cold of the stratosphere was one such completely new battle condition, and even the electric warming suits worn by the bomber boys did not always suffice to stay the terrible claw of the cold. The relentless daily experience of battle, of sortieing day after day without a break, was another.

On a ship, naval battles were separated by quiet weeks of searching for the enemy, steaming across the wide oceans to some remote rendezvous. Days, weeks passed in quiet, and battles were intermittent. On land, similarly, armies spent weeks marching, bushwhacking to meet enemy divisions, advancing through occupied land. But fliers went into action on most days and often every day. They faced the grim reaper each dawn, sun after sun—a strain unimaginable to those who have never known it.

The speed of battle was also new. Air battles flashed past at the speed of a .50-caliber slug, passing in a dive or roll at 30, 40 feet per second. Combat at such a lightning pace had no precedent. Ships moved royally over the waves at 25 to 40 knots—30 to 45 miles an hour. Armies marched at a pace of about three miles per hour, and even armor—tanks, mechanized artillery, armored personnel carriers—traveled at land speed. But air combat rushed by at 400 miles per hour.

All these conditions were especially harsh for bomber crews, whose battle wagons could not divert, climb or dive, but had to maintain a steady course for the accuracy of their bomb runs, making them easy prey for Luftwaffe attackers.

In these new maximum pressure limits, the fliers of World War II encountered for the first time a warfare of unrelenting intensity, which brought never-before-contemplated strain to aviators. Men bent and warped under these new forces. As more cases of shock appeared and pilots began to suffer from "operational fatigue," Army Air Forces surgeons looked for ways to respond. The emotional health of airmen would have to be addressed to keep them flying. Men's psyches suffered emotional wounds as terrible as any inflicted by shot and fire. The face of combat strain took many forms, as commonplace as irritability, frequent anger, sleeplessness, depression, excessive use of alcohol, or far more desperate symptoms.

Major Mort Harris—pilot of the B-17 *Able Mabel* and commanding officer of the 334th Bomb Squadron, 95th Bomb Group—was shot down twice and both times had to ditch in the sea. He recalled one of his crewmen staggering out of their Flying Fortress on return to base, sitting down on the ground and pulling up handfuls of grass. "Well, he sat down and started to eat grass. In front of 150 guys," Harris recalled. "He went nuts."

Harris himself never considered refusing to fly. "I didn't have the guts." But there was talk of mutiny. One crew marched into base and announced as one that they would never crawl into a bomber again. They were transferred, busted in rank and assigned to ground duty. Some men shook uncontrollably on missions. Some became deranged weeks or months after a traumatic event. Others became irrational even during missions.

A copilot of the 100th Bomb Group, the "Bloody Hundredth," worn past reason and responsibility, twice went berserk on missions. In Donald Miller's masterful chronicle of the bomber war, one crewman recalled: "Altogether, we started out for Berlin seven times. Twice our copilot went nuts and tried to crash us into the sea. These times the crew fought him off the wheel and we aborted. After the second time he didn't return to our crew. He wasn't a coward; he just couldn't go back to Berlin."

By late 1943, air force doctors had learned that a key factor in aviators' determination to keep on flying was the knowledge that their ordeal had an end to it. Colonel Grow pleaded with General Eaker to limit tours of duty to 15 missions. Eaker refused, but did cap tours at 25 missions, which in March 1944 became 30 missions and, in July 1944, 35 missions.

Then there were the carefree warriors, the glad gladiators, who were excited by combat and even enjoyed it. Lieutenant J. Clifford Moos, the navigator of a B-24 in the 392nd Bomb Group, was one of those who thrived on battle and even found a thrill in it. "It was exciting. I was 20 years old. By the time we were in combat, it was an adventure. I had no thoughts of fear. I thought it was a little exciting. I called myself a war lover because I had all these adventures," Moos, age 95, said in 2019 before his death in 2021.

"I have never been one who feared dying. Even now I say, if I go to sleep and don't wake up, I lived a good life and accomplished a few things," Moos continued.

Others were simply cold-blooded and took battle for what it was: death for you or for the other man.

A British naval historian and reservist, Arnold Hague, describes the approach to battle of the top World War II destroyer ace, Captain Johnnie F. J. Walker of the Royal Navy. Walker had made up a deadly formula for a submarine attack called the Creeping Attack. "It was cold-blooded murder," said Hague about Captain Walker, a onetime boxer. "It was sheer, calculated murder, but it's war. You want to kill the enemy. He worked out how to do it."

Walker's mechanical attacks became famous—and they worked flawlessly. He sank 20 U-boats, the highest score of any Allied commander on the Atlantic.

But the air war placed new, unprecedented kinds of pressures encountered for the first time in battle upon fliers. When a man cracked in combat or showed dangerous signs of coming undone, he was first diagnosed and cared for by a flight surgeon. More severe cases were treated with drugs such as sodium amytal, which induced a deep sleep for a day or two (sometimes crucial in bringing a man back to wellness), and sodium pentathol, a drug used to relax the patient so he would disgorge memories of the traumatic events that had shaken him.

One of the most common prescriptions for "operational fatigue" was several days at one of the network of country manors the CME had set up; here, airmen under stress could unwind and find peace and companionship for a few days away from the shocks of the clash in the skies. These were the "flak farms." They were as close to a vacation as an airman could get.

The flak farms were comfortable countryside retreats, mostly mansions leased by the Royal Air Force to the USAAF. In these sprawling, peaceful country lodges, war-weary crews found respite from battle and enjoyed the welcome company of American Red Cross female staff. The presence of women alone was restorative, and at least one romance flourished in the network of flak farms. There were 15 of these getaways; they were set up to be as remote from combat experience as possible. Good meals awaited visitors; at breakfast Red Cross hostesses helped plan the day, full of activities like a hunt with hounds, skeet shooting, badminton and bicycling.

In the end, of the estimated 225,000 airmen who flew in combat for the Eighth Air Force through the whole war, some 2,100 were grounded for psychiatric disorders. Many more were treated successfully. The incidence of serious emotional injuries was no doubt kept down by the care of the flight surgeons, who healed the hidden psychological wounds of the crewmen so they could return to their planes for yet another mission.

As 1943 wore on, and the bombers continued to fly missions over Germany without escort to the target and back, the terrible losses kept mounting up, seemingly without putting the slightest dent in the faith of the Bomber Mafia. "Under enormous pressure to prove the efficacy of daylight bombing . . . They dropped record amounts of explosives on Germany's heartland . . . with little tangible results . . . other than the staggering number of casualties on the ground and in the air," wrote one journalist.

Arnold was not alone in his insistence that the raids go on. For the Bomber Mafia—Arnold, Spaatz and Eaker—Operation Pointblank was meant to show off the muscle of their new air force, the vehicle for proving the absolute power of aerial attack in this war and all wars to come. Wrote aviation historian Walter Boyne: they were "totally indifferent to the fact that their politics were awash in the blood of the air crews."

"They were true believers in air power," commented the *New York Times*'s Harrison E. Salisbury in his memoir *A London Diary*. "The important thing, as the Eighth Air Force saw it in 1943, was to establish a presence, to prove a doctrine, to stake out a position in public consciousness. If this cost the lives of many fine young men and inflicted no really serious damage on Germany's fighting capability, that was too bad. War was war and people were bound to be killed." So the gunners, navigators, bombardiers and pilots climbed into their B-17 sky arks; they took off day after day; they intruded over Germany through the flak and the waves of Luftwaffe fighters; and many did not come home.

But Ira Eaker was beginning to worry about the bomber campaign. As the bombers headed out over Germany in the spring of 1943 to implement the Combined Bomber Offensive, Eaker was seeing repeated massive losses of his planes, and by March his fleet was so reduced that he suspended all missions over the German heartland. He limited his attacks to safe targets in France, Belgium and the Netherlands, all the time pleading with Arnold to send more B-17s and B-24s.

Arnold, under huge pressure in Washington to prove his air force, responded with his usual impatience that daylight bombing worked. The problem, as he saw it, was simply General Eaker's lack of will to send his bombers out. His attacks on his friend Eaker became vicious and unrelenting, the cables and letters continuing through the summer as his famous hair-trigger temper got the better of him. "My wire was sent to you," he wrote, "to get you to toughen up." He accused Eaker of "not sending enough bombers on missions because of the fear of large losses." In fact, with each unescorted raid Eaker's enormous losses were depleting his bomber ranks more rapidly than they could be replenished. Eaker did not lack the will—he lacked the planes. There was a continuing lag in the buildup of his forces, bombers in particular, which Arnold might have remedied sooner had he not been so convinced of the bombers' invulnerability. From his Pentagon ivory tower, this was something Arnold seemed unable to understand, and so the pressure on Eaker continued. Wrote one of his Pentagon staffers: "Arnold was terribly impatient. . . . [Eaker] was not creating the miracles that Arnold wanted to report. Arnold just never understood what Eaker was up against." Arnold's celebrated impatience and irascible temper, known throughout the

Army Air Forces, caused immense strain on Eaker, who knew he could expect, as he himself put it, "a bouquet one day, a brickbat on the next." Only his "stoic strength and unfailing sense of humor would enable him to maintain his calm assurance," wrote Eaker's biographer.

In July, Eaker finally had enough bombers to begin the Eighth's first big offensive over Germany. Over six days the fleet hit Kiel, Hamburg, Warnemünde, Hanover, Kassel and Oschersleben, dropping more than a thousand tons of bombs. One hundred bombers were lost and a thousand crewmen were killed or captured. But Arnold continued to press him hard to keep going, and in August Eaker ordered the raid Arnold and his Joint Chiefs of Staff had been planning since March: an attack against the ball bearing plant in Schweinfurt, deep inside Bavaria, which made essential parts for the Luftwaffe's planes, and against the Messerschmitt aircraft factory in Regensburg. The missions were combined, and on August 17, two weeks after the Ploești raid, 376 bombers set out in bad weather. In the most savage air battle ever fought up until then, 500 German fighters rose up to defend the factories. Sixty American bombers were shot down for a loss rate of 16 percent. By now it was obvious the Flying Fortresses could not survive over Germany without escort.

Hap Arnold led a vast two-sky war in both Asia and Europe. He presided over the U.S. Army Air Forces in the Pacific from India to Burma to China, all the while wielding the Eighth, Ninth, Twelfth and Fifteenth Air Forces against the Nazis in Europe and the Mediterranean. He not only conceived and built the modern U.S. Air Force; he was a decisive commander. Arnold began his career in the infancy of aviation and, through persistence and ferocity, not only won World War II in the air but created the globe-straddling reach and command of the superpower U.S. Air Force of today.

But on the urgent need for the Mustang—and on the dubious law of bomber theory—General Arnold blundered. In his stubborn adherence to bomber invincibility and his continuing rejection of the one plane in the Allied pipeline that could protect his bombers— the P-51 Merlin Mustang—he made a costly mistake in a dangerous war. It would still be five months before the P-51 Mustang galloped into battle.

Two days after the August Schweinfurt-Regensburg debacle, Churchill and FDR met in Quebec City, Canada, for the highly secret "Quadrant" conference. Landing in Quebec, Churchill had first journeyed south to Hyde Park for a private visit with Roosevelt at his house on the Hudson, where the two men dined on hot dogs and hamburgers and mapped out their agenda for the meeting. He had detoured 400 miles by special train to show his daughter Mary Niagara Falls, flashing his V for Victory sign through the window at farmers in their fields. Arriving in Quebec City, Churchill was billeted at the Château Frontenac, a huge grand railway hotel built in 1893 by the Canadian Pacific Railway in Old Quebec, the city's historic district overlooking the St. Lawrence River.

Most important on the agenda at Quadrant was whether the British would commit to an invasion of France: Operation Overlord. German airpower, all agreed, must be reduced before Overlord could go ahead. The British pointed to Arnold's heavy bomber losses over Germany. At the current rate of loss, they argued, daylight bombing could not work. Overlord would not be feasible. Arnold was under attack at Quadrant. Publicly he insisted that daylight bombing would succeed. But privately he confided doubts to Robert Lovett, the Pentagon assistant secretary for air and Arnold's close friend, who had pushed him to adopt the Merlin Mustang back in

December. "Hap was having a hell of a time hanging on," Lovett recalled. "I think he was beginning to worry about it because the attrition rate was too high."

After the Quadrant meeting concluded, Arnold flew to England. Having defended it at Quadrant, he needed to see what he could do to make daylight bombing work. "On this trip . . . he saw for the first time with his own eyes what a beating the B-17s were taking," says Arnold's biographer. During Arnold's stay, Eaker put together a mission of 312 bombers to Stuttgart in September. Forty-five did not return, for a casualty rate of nearly 15 percent. Arnold was witness to the wholesale destruction of the bombers as he personally watched many limp in to base in tatters that fall day. And yet, as he studied aerial photographs taken by the RAF of damage done to the ball bearing factories in the August raid on Schweinfurt weeks before, he still seemed convinced the losses were worthwhile. And so the raids kept on launching, the bombers kept flying, wave after wave, into the Nazi cutlass.

In October, Bremen was hit again, then Vegesack and Marienberg on the Vistula River, the farthest-flung raid of the war to date. Anklam and Münster were next. All saw heavy losses of aircraft and crew. One veteran outfit, the 100th Bomb Group—the "Bloody Hundredth"—lost 200 men, nearly half its complement. No air war could be won with this rate of attrition in assets and men. One historian estimated that if the Eighth Air Force's losses had continued at the same rate, it would have had to replace its entire flotilla of aircraft 2.5 times each year.

And then, finally, a second raid on Schweinfurt changed this costly calculus in one decisive Armageddon. No one was even vaguely prepared for the disaster that occurred on October 14, 1943.

CHAPTER 14

Black Thursday

They awoke in the dark at 2:30 a.m. and had breakfast—dried eggs, Spam, coffee, toast with orange marmalade—and then all across East Anglia on the cold morning of October 14 the flight crews of the 19 bomber groups slated for the mission drifted toward their briefing rooms. Three thousand men across more than 50 miles of England were briefed by intelligence officers. The bomber briefings were longer than the fighter briefings because they were more complex, with elaborate information about the target. In the smoky haze of one briefing room near Norwich, the pilots, sitting at every angle on wooden benches and chairs under the glaring lights, saw on the big map that the red string shot straight from their base to the heart of Bavaria: Schweinfurt was the target again. Many of them winced. They had hit Schweinfurt on August 17; 36 of the B-17s in their group had been shot down, almost 20 percent of their planes. No one wanted to go back.

They dispersed to prepare for the raid. They dressed for battle:

heavy electrically heated boots and thick fleece-lined jackets—but not much else; insulation added weight. Takeoff was at 7:00 a.m.

The bombers lifted off before the fighters so that the big wide-winged B-17 Flying Fortresses would have a chance to form up. "You get the four engines going, and you hold your foot on the brakes," recalled Mort Harris, the pilot of *Able Mabel*, whose 95th Bomb Group flew in the second Schweinfurt raid. "You pull the stick right back in your stomach, shove the throttles all the way forward, four throttles in your right hand, and you hold the stick back in your lap. Then—you've got your feet on the brakes, and all of a sudden you let go of the brakes, and it's like a rubber band, a big rubber band—you shoot off like this," Harris recalled before his death at 100 years old in 2021. "And there were trees right in front of us. We were only 200 feet high. You really tried to get over those trees immediately. I mean IMMEDIATELY." Twenty-seven tons of airplane, nine tons of bombs, 2,500 rounds of ammunition and 10 men were airborne.

Almost 300 bombers took part in the assault. They formed up in the skies over England, rendezvoused with their fighter escorts, only short-range P-47 Thunderbolts, and finally gathered in the stronghold of the sky into a convoy covering miles and miles of ceiling. Then they were outbound: objective Schweinfurt.

The 100th Bomb Group cleared the English coast at 12:36 p.m. and met up at the English Channel with the Thunderbolts. "We were flying over an undercast but as we penetrated inland, we began to come out from the undercast," recalled Lieutenant Robert L. Hughes, pilot of the Flying Fort *Nine Little Yanks and a Jerk*. "We could hear chatter on the radio from units and their escort ahead of us. They seemed to be drawing [enemy] fighters." The 100th, the 95th and many other formations roared on into occupied France, the weather chilly and wet, flying east-southeast past Holland,

Belgium and in over Germany—Cologne, Frankfurt. They were grouped into combat boxes for the best protection against enemy aircraft in the cloud cover, but their hidden approach did not last long. At Aachen, on the German border, the P-47s turned to go back, their fuel exhausted. The Thunderbolts waggled their wings in farewell, then dropped away; now the big heavies would continue into enemy territory unprotected, open and vulnerable to the masses of Luftwaffe pursuit planes they expected.

For the fighter pilots, this was one of the most rending moments of their job, leaving the big Fortresses on their own, no longer able to lend the B-17s the power and protection of their eight Browning M2 "Ma Deuce" .50-caliber machine guns. For the bomber pilots, this moment had what the services call "pucker factor," a solemnity that hardened every crewman in the 12 positions that bristled from a bomber's fuselage. There were no more fighter escorts: whatever they faced, they faced alone, with their 12 onboard machine guns. The big, heavy Forts were on their own now, uncovered. The Nazi fighters pounced.

"Once we were on our own, we were under constant attack," recalled one airman of the 333rd Bomb Squadron. The sky turned into a cataclysm of smoke, fire and careening aircraft, the American bombers splayed-out like in a shooting gallery for the German fighter planes. Rockets slammed in from Junkers Ju 88 night fighters. Me 109 Messerschmitts swooped through the boxes of bombers in attack. Focke-Wulf Fw 190s charged directly at oncoming American Forts.

The B-17 Flying Fortress *Cavalier* of the 367th Bomb Squadron, 306th Bombardment Group, was the lone plane of its squadron to hit the target and make it back to base. George G. Roberts, the bomber's radio operator/gunner, recalled the ferocity of the Luftwaffe's interception in the sky-high ceiling. "I had never seen this many fighters before and their plan had been well designed," he remembered. "Twin engine fighters lobbed rockets at the front and rear of our two groups, knocking us out of a tight defensive formation. The single engine Fw 190s and Me 109s then began taking aim at the straggling bombers. The majority lined up at six o'clock high, and, two or three at a time, made a deep dive and began firing their cannons and machine guns. When they got close, they did a barrel roll and peeled off to the right or left, gained altitude and prepared for a new charge."

Now the entire formation was under attack. The Luftwaffe had summoned virtually its full strength to greet the incoming raiders: some 350 fighters, fighter-bombers and night fighters now veered through the formations of American heavies. Flames erupted; flak bursts puffed in the air.

"They knew where we were, and what elevation we were at, and if they hit you in the wing, you'd see a little yellow thing come out of the wing, and that was fire," recalled Mort Harris of the *Able Mabel*. "And if it started, you're dead. It just looked like a little yellow flame a foot high; then it got to be two feet high and about four feet wide and you'd look out and tell yourself, 'I'm gonna last five minutes, maybe two minutes,'" said Harris, who was the squadron commander of the 334th Bomb Squadron of the 95th Bomb Group.

On one mission, Harris had an experience that knocked him mute. One of his buddies was flying as his wingman, a close friend who had joined the Army Air Forces with him, also from southern Michigan. Harris and his pal had trained together, become B-17 pilots together, joined the same bomber group. His friend's plane was hit: "I saw that little yellow flame go up. I knew where he was hit. I was looking out my right window, and he said [through his radio], 'So long, Mort.' And he blew up right in my face."

The thought that jumped into Harris's mind: "There is no God."

Harris had grown up an observant Jew throughout his youth in a suburb of Detroit.

He was in the sky airborne, with no wingman on his right anymore. He was in the command seat of his Fortress with a welter of dials, electric switches, meters and gauges before him on the instrument panel, in his place as skipper. He thundered on. "We just liked each other like brothers," he said of his friend. "Could he play the piano, though, boy. We became good pals," Harris recalled. The bombers continued to target through the vacant expanse filled only with altitude and sunlight and distance. They maintained course.

The battle of Schweinfurt shook and reverberated across miles and miles of sky, across a front 800 miles long, lasting from its beginnings over Germany to its end later in the day. Technical Sergeant Kenneth E. Fox, serving as engineer and top ball turret gunner on a Flying Fortress in the "Fightin' Bitin'" 369th Bomb Squad, fired at incoming planes with the aim of a laser and the power of artillery. In the next moment he was hit; his left side was blown open and his left leg wounded. Fox stayed at his post and kept firing against the many attacking aircraft, not abandoning his position despite excruciating pain and massive blood loss through a 1.5-hour battle, until his plane and crew mates were out of danger.

Other planes also had remarkable stories of duty and destruction. Flak from the ground was intense, concussing all the air around. Lieutenant Hughes in the 100th Bomb Group bomber *Nine Little Yanks and a Jerk* was bringing his ship onto the target, flying it straight down, when a round of flak hit the group leader of the 95th Bomb Group. In the next instant he heard the "WHUMP" of a round hit his wingman, lift the plane up and send it hurtling toward his bomber. Hughes in split-second reflex pulled off a kick of the left rudder, down stick, left aileron, back stick and rolled out of a well-executed diving veer, then peeled away. His wingman was able to thrust his plane through the space Hughes had vacated. Both bombers survived, free in the empty sky. The formation was clear, danger passed. On to the target.

On board *Cavalier*, radio operator George Roberts was in the midst of a nightmare. "Our plane shook from a 20-mm shell that exploded in the waist section and left a hole big enough for a person to fall through, and struck the left gunner in the thigh. The right gunner administered a shot of morphine to the injured airman and we wrapped him in a blanket." Roberts continued: "As I turned back to my gun, a 20-mm shell exploded in the radio room, smashed the communications set on my desk and knocked out the oxygen system." The crew pulled out emergency oxygen canisters. "The other gun positions had only a few rounds left . . . and since we were now able to offer only token resistance, the pilot took evasive action in an effort to throw off the aim of the fighters."

Planes went down all around them. The parachutes of airmen bailing out blossomed like white orchids sailing down the great blue dome of the sky. You could count them: one, two, three; 10 meant an entire crew had gotten out.

Elsewhere the battle played out in thunder, smoke and flame, with amazing luck for some, bad poker for others. A bomber from the 333rd Bomb Squadron drew a tough hand of cards. "One by one our gunners called in saying that their guns were out, so by the time we reached Schweinfurt, dropped our bombs and headed for home we didn't have too much firepower left," a crewman would write later. Then a reverberation. "I'm hit, I'm hit," Dick, another crewman, yelled out. Others were also seriously wounded. Radioman Lou Koth was missing an arm. Still more grave, the aircraft was finished, difficult to fly and losing altitude. Some of the engines had been knocked out. The oxygen system had been destroyed. It was the end of the line; the lady Fortress could crash at any moment behind enemy lines.

They hedgehopped their way at low altitude toward England in their stricken B-17, not knowing where they were, hoping to stay clear of the Luftwaffe and possible attack. Then they lost another engine: the game was up. The pilot picked out a farm with cows grazing in an open field; airspeed was about 150 miles per hour and they were losing altitude fast. The pilot descended and set down, and the plane broke in half. It skidded across a field and into woods, then groaned to a smoking stop.

They were alone, far behind enemy lines, surrounded on all sides by miles and miles of hostile territory. They did not know at first where they were—France or Germany. They had been told back at base, before the raid, that if they were shot down, they should split into small groups and head over the Pyrenees to friendly territory. The unnamed airman and a buddy evaded capture with the help of French resistance agents. They made it back to England within weeks. Others weren't so lucky.

Back on *Cavalier*, George Roberts believed the way out was as perilous as the way in. The German fighters were expected to strike again. One engine was out, another had low power and the only guns left with ammunition were in the navigator's compartment and the radio room. "It was then time for my mission prayer: 'Oh Lord, get me back from this one and I will never go on another.'" No new fighters came up, though; those that remained made only sporadic attacks. "Limping back to the Channel, we encountered total cloud cover and broke off on our own to go down through the overcast, which was so dense that at times it was not possible to see the end of the wing," Roberts recalled. "Following a safe touch down we taxied to our hardstand."

Nine Little Yanks and a Jerk came in just as sure and steady. Lieutenant Hughes reached the English coast at 5:24 p.m. At 6:11 he parted ways with the 95th Bomb Group and brought his plane and his men in to touch down at Thorpe Abbotts at 6:17 p.m.

Only five aircraft of *Cavalier*'s 367th Bomb Squadron had reached Schweinfurt; 10 others were shot down and three had to turn back. The lead group of the formation lost six of 19 planes, and their own low-flying formation lost 13 of 16 aircraft.

Second Schweinfurt was the darkest depth of a dark season for the Eighth Air Force, the low point of the air war. Some 290 bombers had set out for Schweinfurt; 60 had aborted; 229 had gone on to hit the target. Of those, 60 had been shot down, a 26.5 percent loss rate. Seventeen more were scrapped on return; 121 more were damaged. Six hundred forty-two men lost their lives in the raid out of a total on the mission of 2,900 aviators, more than 18 percent of the force. The morale of fliers was at a new low. There was talk of refusing to fly.

Like the Battle of Britain had been before, and then the closing of the Nazi vise on the icy North Atlantic, Schweinfurt was among the grimmest passages in all the Allies' fortunes. The Ploești raid had been known as Black Sunday; now the second Schweinfurt raid would go down in the collective memory of Eighth Air Force bomber crews as "Black Thursday."

Second Schweinfurt was punctuation like a funeral. Everything came to a halt. The Eighth Air Force would not fly again for four months, until February 1944. The generals blamed the halt on inclement weather. And it was true that, as Arnold recorded in his memoir, "In mid-October the weather shut down . . . on southeast

Germany for most of the remainder of the year." But weather or no, it was past time to take stock. At last, Hap Arnold had to turn and face the scrawled signature of reality. He could no longer think of the bombers as invincible.

But, in fact, he had been coming around for months.

The initial key turning the tumblers in the lock was Robert Lovett, the vastly influential assistant secretary of war for air. "Tall, thin, bald, aquiline of face and elegant of dress, Lovett looked a bit forbidding but had wisdom, warmth and wit that won everyone at once," wrote Lovett's biographer. The patrician son of a wealthy family, educated at Yale and Harvard, in the First World War Lovett had formed the legendary First Yale Unit of the Naval Reserve Flying Corps and flown with the British Royal Navy and then with the U.S. Navy, becoming a lieutenant commander. Though he did not look it, Lovett was an experienced pilot who had won the Navy Cross for bravery. He was also a friend of Tommy Hitchcock: Tommy had flown for France in the Great War while Lovett was flying for England.

After World War I, Lovett went to work on Wall Street. On a business trip to Italy, he met a pair of Germans in uniform in the hotel bar. They were Erhard Milch and Ernst Udet, two famous German aces from the early days of Hermann Göring's air force. Their comments about "the Luftwaffe, their country's aircraft industry, and all the things German flying could, and soon would, accomplish," wrote Lovett's biographer, were deeply disturbing to the former pilot. What he heard made him doubt that the American aircraft industry was ready for war with Germany. When he was invited to join FDR's War Department in 1940, he immediately

Edgar Schmued, the designer of the P-51 Mustang. He had a lifelong obsession with flight dating from boyhood; over many years he had conceived of the fighter plane which would turn out to be central to the war. *(Wiki Commons)*

James H. "Dutch" Kindelberger, left, with General of the Air Force Henry H. "Hap" Arnold, was the head of North American Aviation, Inc., which developed and built the P-51 Mustang. A believer in Edgar Schmued's ideas, he led the production of the fighter and became the largest manufacturer of warplanes in World War II. The son of an iron molder from Wheeling, West Virginia, he had started as a draftsman for the Army Air Forces. *(San Diego Air and Space Museum)*

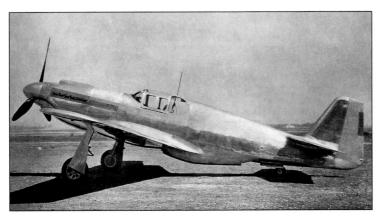

The first prototype of the P-51A Mustang, the NA-73X, in clean Duralumin without any of her markings or colors. She aced her first tests in October 1940. *(San Diego Air and Space Museum)*

Supercharged with the Rolls-Royce Merlin engine, the P-51B was the breakthrough model of the Mustang which turned the plane into the pivotal weapon of the war in World War II. *(Alamy Stock Photo)*

The P-51D was the best expression of the P-51 types. The most numerous production model of the Mustang, the "D" did the lion's share of duty in decimating the Luftwaffe. *(Wikipedia)*

The Merlin 61 engine was the powerhouse that brought the Mustang out of the shadows to the front lines. It had 14,000 separate parts and developed 1,490 horsepower. With Schmued's aerodynamics and the Merlin's thrust, the Mustang became the dragonslayer which eliminated the Luftwaffe from the skies over Europe. *(Author's collection)*

Ronnie Harker, the spirited young test pilot at Rolls-Royce who vaulted the P-51 Mustang to victory. In one prophetic solo test flight, Harker grasped that the underpowered Mustang could become the best fighter plane of the war with the added thrust of Rolls-Royce's Merlin engine. *(Rolls-Royce plc)*

Lt. Col. Tommy Hitchcock, World War I fighter pilot, world famous polo star, investment banker, and air attaché at the American Embassy in London. He would become the leader of the drive to mass produce the P-51 Mustang in America. *(Courtesy of Louise Hitchcock Stephaich)*

Tommy Hitchcock in uniform around the time of his duty with the French Air Service in World War I. He scored two victories over enemy planes and received the Croix de Guerre with two palms. Shot down by three German adversaries, he later escaped from a prison camp. *(Library of Congress)*

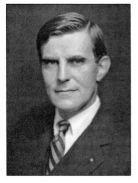

John G. Winant, U.S. Ambassador to London during the war, worked closely with Lt. Col. Tommy Hitchcock, his assistant air attaché, to publicize and bring to the attention of the Pentagon the remarkable P-51B Mustang with its outstanding capabilities. He and Hitchcock were the vanguard of a colony of British-based government and military officials in the U.S. Army Air Forces and the British Royal Air Force militating for adoption and mass production of the Mustang. *(Library of Congress)*

Robert A. Lovett, who would rise to become Secretary of Defense, was one of the first Army Air Forces officials to understand how vital the concept of a long-range fighter aircraft—and the P-51 Mustang—were to the air war in Europe. After a fact-finding trip to England in 1943, he pressed Army Air Forces chieftain General Henry H. "Hap" Arnold to implement fighter escort using the P-51 Mustang. A fighter pilot in World War I, Lovett was a close friend of Tommy Hitchcock and American Ambassador Gil Winant. *(trumanlibrary.gov)*

Col. Donald J. M. Blakeslee commanded the 4th Fighter Group during the war. He led the way in bringing the Mustang to the front lines of the Eighth Air Force in Europe. He was a hard-charging fighter ace and a great leader; it was said he could drink all night and fly skillfully all day. *(U.S. Air Force)*

The P-38 Lightning preceded the Mustang and had good range but a lot of engine problems. It was far more successful in the Pacific than in the cold of the skies over Europe. *(Wikipedia)*

The Republic P-47 Thunderbolt served early in the war and also preceded the Mustang but had a short range and poor rate of climb. Many pilots still praised its sturdiness, stability, and diving performance. *(Alamy Stock Photo)*

The Boeing B-17 heavy bomber was the centerpiece of the Allied air war. It could carry nine tons of bombs and was defended with 12 machine guns, but was no match for the German Luftwaffe fighters which ravaged unshielded bomber forces. *(U.S. Air Force)*

The Consolidated B-24 heavy bomber was the second heavy bomber flown by the U.S. Eighth Air Force. It could carry a five-ton payload of bombs, and it hit, among many other targets, Ploesti, Schweinfirt, and Berlin. The B-24 mounted ten .50-caliber machine guns and had a speed of 290 miles per hour developed by four Pratt & Whitney radial engines. *(Wikipedia)*

Col. Joseph A. Peterburs, then a lieutenant, nicknamed "The Strafing Kid." He destroyed five enemy aircraft on the ground and shot down top Nazi ace Walter Schuck, the winner of 206 aerial duels, in an Me 262 jet. The two later met and became friends. *(Courtesy of Joseph A. Peterburs)*

Col. Wallace E. "Lucky" Lowman, of the 55th Fighter Squadron, 20th Fighter Group, who theorized that most fighter pilots felt there was "a job to do, and where do we go?" Lowman went on to become a career Air Force officer. *(Author's collection)*

Lt. William "Mac" MacClarence who helped bag one of the experimental Luftwaffe "Swallow" jets. He called being a fighter pilot a lonely job in which you prayed not to foul up. *(John Dibbs, planepicture.com)*

General Charles E. McGee as a young pilot in the all-Black, segregated 332nd Fighter Group, the "Tuskegee Airmen." They became one of the most successful and sought after escort groups for bombers going into Germany. They flew 179 escort missions and lost bombers on only seven of them. *(National World War II Museum)*

Capt. Thomas M. Stanback, Jr. was the statistical officer for the 20th Fighter Group. He admired the exploits of the pilots and remembered vividly the Group Commanding Officers who led their men into battle on missions in P-38s and later P-51s. *(Author's collection)*

Many years after Col. Joe Peterburs shot down Nazi ace Walter Schuck, who claimed 206 victories, the two met and became fast friends. They are pictured here at a reunion. *(Joseph A. Peterburs)*

Cliff Moos, left, with his brother, Jerome; they were both assigned to the same Bomber Group, the 392nd, based in Shipton, England. Cliff Moos thought his war as a 1st Lieutenant and navigator in a B-24 Liberator was an adventure and often exciting. *(Courtesy of Jane M. Cohen)*

The Messerschmitt Me 109 was a standard of the German Luftwaffe. It had a top speed of 386 miles per hour and was powerfully armed with one 20-mm cannon and two .50-caliber machine guns. It was the best known of the German fighters, having first flown in 1935. *(Alamy Stock Photo)*

When it appeared in battle in the summer of 1941, the Focke-Wulf Fw 190 outclassed all Allied aircraft. Called by German pilots *der Würger*—the "Butcher Bird"—the Fw 190 had a top speed of 406 miles per hour and could climb 19,685 feet in just under ten minutes, or about 36 feet each second. It raised alarms among British Royal Air Force leaders. *(Alamy Stock Photo)*

Another deadly bird of prey from Messerschmitt, the Me 110 was a two-engine night fighter which showed its effectiveness against undefended bombers. It packed heavy firepower, two 30-mm cannon in the nose, two forward-firing 20-mm cannon under the belly and one twin-barrel machine gun in the rear cockpit. It was designed as a *Zerstörer* aircraft, a "Destroyer." *(Bundesarchiv)*

A B-24 over Ploesti, Romania. The facilities at Ploesti provided 60 percent of Germany's crude oil and the Allies made numerous raids on the installations. The first on August 1, 1943 was a tough mission without fighter escort on which bombers from the Eighth and Ninth Air Forces pounded the target but sustained heavy losses. *(U.S. Air Force)*

The B-17 heavy bomber *Lady Bedlam* and a flotilla of Flying Fortresses on a raid over Germany. The Forts had a top speed of nearly 300 miles per hour and a service ceiling of 35,600 feet. When accompanied all the way to targets inside Germany and all the way home by Mustang long range fighter escorts, these bombers ravaged Nazi industry and cities and brought an end to the war in Europe. *(U.S. National Archives)*

General of the Air Force Henry H. "Hap" Arnold built the modern U.S. Army Air Forces to superpower status and led the victorious world war effort in the Pacific and Europe. He had learned to fly with the Wright brothers and sang the praises of U.S. bombers, but, critically, he failed to appreciate the urgent need for fighter escorts to shield them. He ignored the P-51 Mustang and refused to accept the clear need for a long-range fighter plane until late 1943. Air Force historians called it "the costliest mistake made by the AAF in World War II." *(National Museum of the United States Air Force)*

Lieutenant General Ira C. Eaker was the leader of the U.S. Eighth Air Force, the American aerial presence in Europe, from its origins in February 1942. When Gen. Arnold replaced him in early 1944, Eaker became commander-in-chief of the Mediterranean Allied Air Forces. *(U.S. Air Force)*

General Carl A. "Tooey" Spaatz was one of the original "Bomber Mafia" who fervently believed in the invincibility of bombers. As commander of the U.S. Strategic Air Forces in Europe in 1944, he argued vociferously for concentration of Allied bombing on oil facilities, helping to cripple Nazi Germany's ability to make war. *(U.S. Air Force)*

General of the Air Force James H. "Jimmy" Doolittle, former racing pilot, speed record holder, test pilot and leader of the famous first ever raid on Tokyo in 1942, took over command of the U.S. Eighth Air Force in January 1944 and led its final drive to victory. He instituted "freelance" fighter tactics, freeing fighter escorts from staying close to bombers and releasing them to go on hunter-killer assaults on Luftwaffe fighters. *(U.S. Air Force)*

Lieutenant General William E. Kepner, right, with Col. Don Blakeslee after the Russia Shuttle, was a pioneering balloonist who served in the infantry in the First World War. General Kepner became chief of Eighth Air Force Fighter Command in September 1943 and commanded Eighth fighters which did so much to destroy German forces in preparation for D-Day. He himself flew 24 combat missions in fighters and bombers and was awarded the Distinguished Service Cross and numerous other decorations. *(American Air Museum in Britain)*

accepted the job in order to help American aviation prepare for the war he knew was coming.

In May 1943, concerned about daylight-bombing losses as the Combined Bomber Offensive got underway, Lovett flew to England to size up the situation. In London, his old friend Hitchcock entertained him. Now returned from Washington, Tommy had taken a house near Grosvenor Square and the American embassy, where he continued to entertain to further the cause of the P-51. This particular evening he gave a dinner for Lovett to meet the top RAF brass. The next day, he sent the place cards from the dinner to Peggy in New York. The guest list included every important officer in the RAF and a smattering of American dignitaries as well.

From London Lovett went on to the headquarters of the Eighth Air Force outside the capital. He was met by General Eaker, who had by now assumed command of the Eighth, and stayed with him at Castle Coombe, an elegant Tudor house with gardens, a tennis court and a bomb shelter that was Eaker's home nearby. Already friends, Lovett and Eaker were able to talk in depth about the daylight-bombing situation in Europe. The party of Americans visited aircraft factories and Eighth Air Force bases, talking to pilots and ground crew everywhere they went in East Anglia. Lovett learned for himself the truth of the bomber war: that the fighters' limited range forced them to turn back long before reaching the target, leaving the bombers unprotected. The airmen's pleas made a deep impression on him: bombing would never succeed without protection all the way to the target zone and back out to clear skies. They only confirmed his feeling that long-range fighters would be the key to air superiority over the Germans.

As soon as he returned to Washington, Lovett wrote a long memo to Arnold stressing "the immediate need for long-range fighters."

Fighter escort would "have to be provided to B-17s on as many missions as possible . . . in order particularly to get them through the first wave of the German fighter defense which is now put up in such depth that the B-17s are forced to run the gauntlet both into the target and out from it." By then Germany, defeated at Stalingrad and in North Africa, had returned several veteran *Jagdgeschwader* units to Europe: her daytime air defenses were much better than in 1942.

"I pushed hard on Hap," Lovett would recall later. "He said our only need was Flying Fortresses, that's all; [that] very few fighters could keep up with them. . . . The Messerschmitts had no difficulty at all." In mid-June, Arnold had Lovett's memo on his desk. His anger continued to thunder down on his staff in England, particularly his old friend Ira Eaker. He continued to blame Eaker for the Combined Bomber Offensive falling short. But, in fact, Lovett was speaking out loud doubts that the chief of the Army Air Forces was beginning to feel himself. Publicly, Arnold still held the position that the bombers were invincible. But privately he was starting to wonder. With Lovett's memo, obstacles to fighter escort began to fall.

On June 22, Arnold penned a directive to Major General Barney Giles, his director of military requirements. "Within this next six months you have got to get a fighter to protect our bombers. Whether you use an existing type or have to start from scratch is your problem." Lovett's memo to Arnold concluded with the words "High hopes are felt for the P-51 with wing tanks." Arnold did not mention the P-51, but Giles thought of it right away. The Mustang with a Merlin engine, he knew, had the performance to match the German fighters. But it still did not have quite the range for long escort duty. He called Dutch Kindelberger out at North American. He told him the Army Air Forces wanted to put 300 more gallons of gas in the P-51. Dutch was appalled. He was sure, wrote Schmued's biographer, Ray Wagner, "the wings would not be strong enough and neither would the landing gear."

"Have a P-51 on the line tomorrow morning at ten o'clock," Giles said. "I'll fly out there tonight." The next morning, after examining the plane, Giles said, "Take out the radio behind the pilot and put in a 100 gallon tank. Then open up the wings and put in bullet proof tanks the entire length." Standing on the tarmac at Inglewood that June morning, Kindelberger and Ed Schmued "didn't believe the plane could take such a load." But Giles insisted. "Put the tanks in . . . and fill them with water and give the plane a thorough test." Within weeks "the water-laden P-51 underwent and passed a series of tests that would soon have a profound effect" on the air war in Europe.

Then Bob Chilton, the pilot who had first tested the converted Merlin Mustang for North American the year before, was directed to simulate a flight to Berlin before the aft tank was approved. He needed to prove that the P-51 with the new tanks would have the range for long missions over Germany. Chilton recounted the ordeal: "I ran out of gas before the airplane did. . . . I was sitting there seven and a half hours. It was horrible. . . . I went up and down California . . . San Diego . . . Santa Barbara . . . seven and a half hours was brutal in an airplane."

At Giles's urging, Arnold ordered additional Mustangs for delivery to England. But they could not possibly get there until late fall— and the war would not wait. The Eighth Air Force would have to fight the Germans with what it had. In the last week of July, the Eighth dropped more than a thousand tons of bombs on Germany. But they also lost 100 bombers, and 1,000 crewmen were killed or captured. And then on August 17 the doomed first raid on Schweinfurt and Regensburg ended in disaster.

By the second Schweinfurt raid in October, even members of Congress were agitating. An outcry was raised on Capitol Hill. It seemed to many that Arnold, "in his Washington ivory tower, did not have the remotest idea of the extraordinary physical and emotional costs exacted by all-out air war." In a press conference the day after the raid, FDR hedged in his response to questions from horrified reporters about the loss of 60 bombers in a single day.

But it was too late. The press began to protest the direction of the air war. There was spreading discussion of mutiny. As one historian wrote, the air force brass now knew "that their basic operational assumption was faulty: the bombers could not get through without fighter protection, and the existing fighters were too short-legged."

The notion of bomber invincibility was in ruins. Arnold ordered full steam ahead on Mustang production. In late 1943, Mustangs began to roll out of North American's Dallas and Los Angeles factories. What had begun as a trickle of 70 planes in March 1943 became a surge: 295 in November, 332 in December, 370 in January 1944, 380 in February, 482 in March, and 700 planes a month by August 1944. By 1944 the North American factories were spewing out P-51Bs like so many M1 Garand rifles. Soon hundreds, then thousands of new Mustangs were being flown to the East Coast. They crossed the Atlantic on the decks of convoy ships or were flown to England. Help was on the way.

Tommy Hitchcock's tireless crusade had prevailed at last, but his victory had come at the eleventh hour. D-Day was looming less than six months away. The Luftwaffe dominated the skies, with Nazi aircraft factories producing 3,400 Fw 190s, over 6,200 Me 109s and some 1,500 Me 110s—a total of more than 24,800 planes of all types—in 1943 alone. Hitler, Göring and their *Luftflotten* were still masters of the air.

PART THREE

HAWK OF HEAVEN

Don Blakeslee

CHAPTER 15

A Maverick Fighter Ace

He was a six-footer, with a powerful, athletic body that seemed always on the verge of lunging, jumping or uncoiling; he had leaden blue eyes and a squarish face that looked like it would never betray any emotion, whatever the situation, whatever the time, and in the bright morning sun the figure etched against the sky was recognizably American. Early in the cold on December 1, 1943, Lieutenant Colonel Donald J. M. Blakeslee climbed out of a beer-barrel Republic P-47 Thunderbolt, raffishly handsome, with wavy dark hair, fixing the ground crew with his penetrating gaze. As the most experienced fighter pilot in the U.S. Army Air Forces in England, veteran of 130 sweeps with the RAF and a year of missions with the Eagle Squadrons, Blakeslee had been seconded from his own 4th Fighter Group to the 354th to break in the pilots as they switched over from P-47s to the brand-new Merlin Mustangs. The first P-51Bs would take wing under Blakeslee's hand.

All through the fall, with the Army Air Forces finally embracing the Mustang, a flood of P-51s had streamed, then surged from North American's factories in Los Angeles and Dallas and been flown cross-country by the women pilots of the newly formed WASP Ferry Command. Sometimes the last leg across the ocean would be by ship, but often the WASPs would fly the Mustangs across themselves.

There was great excitement on the waiting bases when the new Merlin Mustangs arrived. As the new planes landed, "the engines shut down, and the canopies opened," the pilots stepped onto the tarmac. Often when they "pulled off their leather helmets and shook their heads, their long, feminine blond and brunette hair swirled in the breeze." They were women, and they had flown across the Atlantic. The men on the fighter bases could not believe their eyes.

The 354th Fighter Group, part of the Ninth Air Force, had received the first Merlin Mustangs in December; it was the first of many outfits to be equipped with the remarkable new plane. There was one last hurdle, though, for the bird that had traveled on such a long pilgrimage across almost four years, from the drawing table of Edgar Schmued to the airfields of southern England. General Eaker and the air force had assigned the new fighters to the Ninth for tactical operations, support duty—not to the Eighth for the frontline high-altitude bomber-escort work over Germany for which they were so desperately needed and so exquisitely suited. It seemed the Army Air Forces still did not understand what the P-51B could do.

The arrival of the Mustangs caused a sensation among the young pilots at the small airfield in Boxted, between Cambridge and London. Don Blakeslee would be their master and coach, leading them on their first missions flying the P-51B. In the end, he would do more than any other officer in the USAAF to bring the Mustang to the 200,000 personnel, 14 fighter groups and 40 bomber groups of the now enormous Eighth Air Force. Blakeslee would become the apostle of the P-51, proselytizing and pushing for it everywhere. As Tommy Hitchcock had taken up the cause in London and at the Pentagon, Blakeslee would carry it to the Army Air Forces brass in England.

An Ohioan from the small town of Fairport Harbor near Lake Erie, Blakeslee was the kind of fighter pilot who had lost his heart and head to flying when still a youth; throughout his life, his obsession with flying would border on the addictive. Rising from a hardscrabble background—his working-class parents were 19 when they married, and his father soon abandoned the family—he had fallen in love with airplanes watching the Cleveland Air Races as a boy. With money saved from a job at the Diamond Alkali Company, he and a friend bought a Piper J3 in 1939. When his friend crashed the plane, Blakeslee decided to join the Royal Canadian Air Force as his best chance to keep flying. He loved training in Canada, flying far out over the empty plains, buzzing foxes in his aircraft.

Once in the RCAF, Blakeslee, like Chesley Peterson, had joined the Eagle Squadrons, three fighter squadrons made up of some 240 American volunteers—adventure seeking, patriotic or both—who made their way to England to fight in Europe's struggle before the U.S. joined the war. With America still neutral, they often had to evade the FBI to cross the Canadian border and sign up. Like Blakeslee, many of them were boys who loved to fly but did not have the college education or prior military experience required to enlist in the U.S. Army Air Corps. They had only good eyesight, a high school diploma and tremendous courage: 100 of them would die in action by the fall of 1942. Arriving in England in the spring of 1941, Blakeslee had scored his first victory in November—a German Me 109—and then, in quick succession, won the British Distinguished Flying Cross and achieved ace status, chalking up five kills by August. He was 25 years old.

That fall, still flying Spitfires, the three Eagle Squadrons were transferred to the Eighth Air Force as the 4th Fighter Group. But Blakeslee, by then one of the half dozen most highly esteemed pilots among the Eagle volunteers, had been "busted" for entertaining female company in his quarters. One night an elderly group captain making the rounds had seen not one but two WAAFs jump out the window of Blakeslee's room. The American general reviewing Blakeslee's case was amused. "Two women?" he said. "And he's up for major? Hell, I'll make him a colonel!"

A hard-drinking, soft-talking, doggedly aggressive pilot, he was a raging bull of a man of contrasting quiet and furor who seemed to be able to romp all night and fly all day. Like a character out of Hollywood central casting, he caroused, womanized, swore frequently and often lost his temper. At times cool and aloof, at others charming, he was a hard man to get to know. But he was, as one admirer put it, "probably the best [air] leader, one of the finest pilots around." Blakeslee could be shy on the ground, but in the air he was a fierce leader with a ferocious attack mentality. He was "ruthless . . . relatively fearless and would never miss an opportunity to fight the enemy in the air," observed one flier. "In the air he was all business, and the business was killing."

That December in the briefing room at Boxted, he told the 354th's young pilots that it would be better not to return from a mission than to break radio silence. And they must never—repeat, never—turn aside from the enemy. Leading the first sorties in Merlin Mustangs, he came to a decisive conclusion up in the clouds: the P-51B was the best fighter he had ever flown. That conviction only mounted as he led the 354th in combat. From the outset, the virtuosity of the new bird was dramatic. So, as the weeks went on, was its battle record.

On January 4 over Kiel, 14 Messerschmitts were shot down in combat without loss to the 354th. Seven days later, in a raid over

Halberstadt and Oschersleben, a Mustang of the 354th piloted by Major James Howard engaged German fighters attacking a formation of Flying Fortresses for fully half an hour. Despite three of his guns jamming, Howard, diving and veering in his new Mustang, shot down six enemy planes, charging and attacking until his fuel ran low. Sixty Forts were lost on that January 11 mission, but not one from the group Howard had been defending.

The breakthrough formula had been found: a fighter with long enough legs to escort the bombers into Germany and enough speed and maneuverability to outfly the Luftwaffe. Expecting vulnerable bombers without escorts, German pilots were stunned to encounter American fighters intruding over their homeland, penetrating so far inside Germany. The day of reckoning had arrived.

With each new mission, Blakeslee's certainty about the P-51 only hardened. Each night he could not resist flying back from Boxted to his own base at Debden, parking his Mustang and showing it off as wide-eyed young P-47 pilots crowded around to admire it and hear Blakeslee's stories of its feats in combat. "It's the ship," he would say, using pilots' slang for an aircraft. "It's the ship," the awestruck young pilots would repeat. The Mustang was what the boys at Debden liked to call a "sweet" airplane.

Blakeslee had no love at all for the Republic P-47 Thunderbolt. He had flown Spitfires with the Eagle Squadrons and loved their streamlined speed and their acrobatics in the air. To him, the P-47 was a lumbering crate that couldn't climb. Even more important, he had seen firsthand the effect of its limited range. Since the fall, Blakeslee's 4th Fighter Group had been the only group in the Eighth flying escort with the P-47. He had hated the feeling of having to turn back at the German border, leaving the bombers to continue on over Germany to certain slaughter.

Now, flushed with first exposure to the agile new fighter, his pursuit of the P-51B for his own 4th Fighter Group and then for the whole Eighth Air Force became an idea, then a passion and then a burning quest. Blakeslee began to lobby his superiors, buttonholing and badgering them personally, but to no avail. All P-51s continued to be assigned to the Ninth Air Force for tactical work. Finally Blakeslee determined he would take his case straight to the top—to Major General William E. Kepner, by then the head of all the Eighth Air Force Fighter Command, directly at his headquarters in Honington, Bury St. Edmunds, Suffolk. The young pilot pleaded with the general to outfit the Eighth with Mustangs. Kepner re-fused. A fan of the maverick young colonel, Kepner was sympa-thetic to Blakeslee's cause, but a great aerial offensive was in progress. The Luftwaffe must be brought to heel before D-Day. This was no time to switch machines. Blakeslee screwed his argu-ments tighter. Kepner rejected them. It would take days, Kepner said, to retrain pilots and ground crews on the new plane.

"No sir, General," Blakeslee retorted. "Most of these boys flew liquid-cooled types for the RAF. It won't take them long. As for the mechanics, don't forget they worked on Spitfires when the group first started to operate. Don't worry about them." Kepner hesitated. "General," Blakeslee implored, "give me those Mustangs and I give you my word I'll have them in combat in 24 hours. I promise—24 hours." Kepner came through with the Mustangs a few days later.

To the pilots Blakeslee said, "Don't worry, you can learn to fly 'em on the way to the target." And they did. In the end the 4th's pilots, who normally averaged 200 hours of training on a new air-craft, got some forty minutes on the Mustang before they flew their first mission. The night the new Mustangs arrived at Debden, the excitement was barely contained. "You couldn't tell enlisted men from pilots that raw night of February 27, 1944, for they were both doing the same thing: washing, gassing, and slipping the snaky .50-cal armor piercing incendiaries into the wing guns," wrote Grover Hall in his memoir of the 4th Fighter Group. The pilots didn't care: they were trading their mules for stallions.

From then on, the P-51Bs and Cs (identical but made in Dallas) began to feed across the ocean, out from British ports and through the Eighth Air Force like migrating herds from a winter collection point, then scattered to the corners of the compass. By February three groups had the plane—the 78th at Duxford, the 358th at Leiston, the fabled 4th at Debden; two more groups in March; two more in April, the 359th at East Wretham and the 339th at Fowlmere; fully nine of the fighter groups before D-Day. Finally, by September 1944, all fifteen U.S. Eighth Air Force fighter groups in England, with one exception, would be supplied with Mustangs— some 1,400 planes. Only the 56th, whose illustrious ace Colonel Herbert "Hub" Zemke favored P-47s, would retain them until the end of the war. The Mustang, for all intents and purposes, was now the fighting maw of the Eighth Air Force and the air war in Europe.

In January, General James H. "Jimmy" Doolittle—holder of the 1932 air speed record, test pilot and racing pilot, hero of the stunning 1942 bomber strike on Tokyo—was brought in to replace General Eaker as head of the Eighth Air Force. By now a true believer, Doolittle advocated forcefully for more P-51s in his armada. Then things moved quickly. General Arnold, finally a convert, transferred all Mustangs assigned to the Ninth Air Force to the Eighth and placed them under General Kepner. All U.S. pilots with any Mustang training were ordered to Kepner's squadrons. The RAF turned over four of their Mustang squadrons to the Americans. The winding trail had ended for the NA-73X. It was locked and loaded, the most strategically important air weapon in the Allied lineup.

Bad weather in the winter of 1943–44 had forced a pause on bombing missions over Germany. The Eighth Air Force had hung in suspended animation. But toward the end of February the skies cleared and Operation Pointblank resumed. With increasing numbers of Mustangs guarding the huge bomber formations, the Eighth moved ahead to obliterate the Luftwaffe with an intensity that pounded across the German homeland. American and British bombers hit both airframe and aircraft engine production plants, ball bearing and synthetic rubber factories—all the components of the German aircraft industry. And Doolittle's radical new strategy released the fighters from sticking close to the bombers, permitting them to strike out as hunter-killers, chasing down and bagging enemy airplanes ambushing the Forts and Liberators. Released now to "freelance," the P-51s were able to do much more than defend; they could carry the battle to the enemy and hit the marauders in aggressive attack. The results were spectacular, and the new doctrine stuck to the end of the war. The hour of the Mustang had come.

In a massive joint operation—slugged "Big Week"—Allied bombers flew 3,800 sorties and dropped 10,000 tons of bombs that February on German aircraft component and preparation plants—more bombs than the Eighth had dropped in its first year of operations. Only 226 bombers, or six percent, were lost; of 3,700 fighter sorties, including Mustangs, casualties amounted to a mere 28 aircraft, or less than 1 percent. P-51s were registering the first flashes of victory.

The wings had reached the battlefront. The hard-luck Eighth Air Force was finally armed to destroy the Luftwaffe in the stronghold of the skies; at last the battle-weary heavies would have an airborne fighter shield to take them through the surprised Nazi defenses. It was twilight across the horizon, though, and the sky was growing dark. It was just three months until D-Day. The Mustangs of the Eighth had a big job to do and no time at all.

CHAPTER 16

The Jig Was Up

On the chilly, socked-in Saturday morning of March 4, some 65 pilots waited, eager and tense, for any word on the operation they were about to unleash. Many men were smoking; others chewed gum; many were silent. At the front of the dingy briefing room of the 4th Fighter Group base hung a vast map of Europe with a red line in crayon stretching from Debden to Berlin. Never before had they struck so far into the interior of Germany, a 1,200-mile round trip. The heavily defended citadel of the Reich, bombed by the French, the Russians and many times by the British since 1940, had sat untouched by fighter aircraft, raided only by exposed bombers completely unprotected by escort. The Americans themselves had never hit Berlin at all. The Nazi capital had sat unpunished by America through the fall of France, Hitler's advance into Russia, the Nazi U-boat assault on American shipping, the invasion of North Africa. Until this morning.

Now a hush fell as Blakeslee walked into the briefing room

wearing the beat-up leather jacket that was his trademark, a cigarette in his right hand, his pale blue eyes glinting. He strode quickly up to the front of the hall and turned around. "Okay," he said. The men fell quiet.

"Well, you've seen what the show is," he told them. "We're going to Berlin."

The raid was not only unprecedented, it would be titanic, on a scale never before encountered. The clash would bring to battle some 600 Allied heavy bombers and 400 Nazi defenders; it would involve a total of some 600,000 Germans and Americans, including 12,000 Allied airmen in B-17s and fighters and 1,000 Nazi airmen. Nineteen million rounds of machine-gun ammunition and 120,000 rounds of cannon ammunition would be loaded into the Allied planes.

The 4th had received their new Mustangs on February 27. Now, just six days later, Blakeslee would lead the 15-mile-long parade of American bombers and fighters thundering across the heart of Berlin for the first time to cut through the German fighter screen and set fire to the city. His relentless campaign to get Mustangs for the Eighth had earned him the honor of leading the first American strike on Germany's capital.

British orderlies had awakened the pilots at 4:00 a.m. Shaking each man, the "batmen" had said as usual, "Time to get up, sir, nice morning, sir."

Now out at the dispersal hut, the air was charged. No one spoke. Blakeslee looked at his watch, then down at his black flying boots. For a moment he was no longer commanding officer of the base. He was just another young pilot about to undertake a mission from which he might not return. On the runway, the wind was icy. The mechanics shivered in their green coveralls as they waited to see the pilots off on the historic mission. Then the intricate near circus as the Mustangs lifted into the sky and joined the bombers, hundreds of planes gathering in formation and configuring in combat boxes and echelons, B-17s and Mustangs four miles up in the firmament.

The immense armada of wing power made its way at 20,000 feet toward the enemy coast in cold that was 56 degrees below zero. Its vapor trails and lines of aircraft covering several square miles of sky, the formation roared ahead in tight order as it droned east. Then letdown. The weather turned dark; stormy fronts blew in; the massive raid and its giant formation of bombers had to be recalled. The mission was scrubbed and the heavies roamed back to base, landing in the overcast.

Two days later, on March 6, the weather was perfect. The ceiling was ready, and so were the pilots and planes—Forts, Liberators, Mustangs; also Lightnings and Thunderbolts for short-range escort to the German border. Blakeslee was again to lead into battle. Once more, the mammoth flotilla lifted off the runways, linked up in the skies over England and headed across the Channel.

"Those 700 planes, or 500, whatever they decided to put up, created a humongous jet stream," recalled then-Lieutenant William R. MacClarence, a Mustang pilot who flew with the 339th Fighter Group, about big missions like this. "You'd pick up this huge jet stream. You'd know about what time you'd pick up your bomb group. You were aware that when you'd found the bombers, you'd find a great big contrail. And if you didn't pick up your group right away, you flew down along that contrail."

Then they struck out for Germany, a migration of will and superforce, the bombers roaring on in their combat "boxes"—a high group of seven or eight planes, a low group, a lead group and a tail squadron—arrayed in an interlocking defensive grid bringing a total of some 390 bristling guns to bear on the enemy attackers.

For the Mustangs, it was their first starring role: they would charge into action like a division of cavalry driving across the Berlin sprawl, striking enemy fighters from the sky, diving through the formations and echelons of bombers with Browning .50-caliber machine guns smoking. It would be the first time they appeared over Berlin, and the Luftwaffe would be stunned on March 6. The 4th Fighter Group, the 357th, the 55th and the 354th would go on to successfully defend heavy bombers, kicking off an onslaught that would continue month after month until D-Day and beyond. Like the Greeks at Salamis ages before in 480 BC, they would begin to reverse the fortunes of battle. They would begin to steal the skies from the Luftwaffe.

But first there was Berlin, a baptism by fire for the P-51B. The Mustangs swept into the Berlin sky space like a wind to wake the world.

That day, the armada climbed into the atmosphere and winged out over the Channel, the contrails leaving white wakes in the sky. They came in over the coast, fighters and bombers filling miles of sky, droning in on target Berlin.

The Mustangs had names; they almost had personalities. *Josephine* was named for Joe Peterburs's sweetheart and later wife; *Cripes A' Mighty* was the plane of top Mustang ace George E. Preddy, Jr., who had 26.5 aerial victories; there was *Big Beautiful Doll, Happy Jack's Go Buggy, Impatient Virgin, Sizzlin' Liz, Gumpy* and *Man O' War.* The bombers had names, too: *Strawberry Bitch, Lady from Hades, Liberty Belle, Flak Dancer, Belle of the Brawl, The Joker.*

The climb to bombing altitude had been steady, even, continuously ascending in the quiet, bright morning sun. The vast formation passed over the enemy coast near the Hook of Holland at 12:10 p.m., 600 bombers and hundreds of supporting fighters arrayed like a migrating flock or an immense flight of wings thundering in, homing in. The bombers were positioned in their combat boxes, the fighters above and to the sides of the big B-17 and B-24 battlewagons, a huge congregation overhead. Now the whole enormous army in the sky flew in tight configuration over Holland and Jutland, crossing the broad, receding expanse of Nazi-occupied Europe.

The flak hit first. Fire came up from the gun emplacements below almost as soon as the convoy had crossed into Germany. The flak was as lethal as any fighters, but of varying intensity as the Allied flotilla drove ahead. Now they passed from the big North Sea port of Bremen to the Elbe River and Hamburg on a straight easterly line across northeast Germany, then south and east over Wolfsburg, Oranienburg and Hanover. The antiaircraft guns tracked them, shooting high at the advancing fleet overhead. Black puffs filled the ceiling and explosions rocked the thin air at 20,000 feet. Over Vechta, just south of Diepholz, antiaircraft fire was "meager" but "accurate."

Then the Nazi fighters came in; they shot onto the stage and swarmed over the Allied array. Some 400 aircraft were lofted into the air that day by the Luftwaffe over the bastion of Berlin; they came up in waves—50, 75 single-engine fighters, night fighters, bombers, Junkers Ju 88s, Messerschmitt Me 110s. The navigator of an American bomber in the 381st Bomb Group, Lieutenant Charles A. Gilpin, Jr., saw it this way: "We had been over enemy territory about 45 minutes when we were struck by about 75 Fw 190s, Me 110s, Me 210s and Ju 88s. We fought them savagely for about 30 minutes. . . . On top of that, we could see a solid wall of flak waiting for us ahead. We mentally tightened our belts and plowed through." Then, in the intensity of battle, the Mustangs went into action. The P-51Bs lit out, pounding the flights of Nazi furies, cutting them down, unleashing unbroken, long-awaited power.

Lieutenant John T. Godfrey, leading ace from the "Fourth But

First" 4th Fighter Group with 18 kills, spotted more than 12 twin-engine bandits below him and more than 20 single-engine fighters attacking from the side of a bomber. He targeted a Focke-Wulf. "I started closing . . . and gave him a short burst. . . . Again I opened up and this time I saw a couple of strikes on his starboard wing." Godfrey had to let his quarry go because several other Fw 190s were on his tail; before long he came upon an Me 109 head-on and attacked. Only one gun fired, but Godfrey sped on toward havoc anyway. "The Me 109 made three head-on attacks and on the last one I managed to turn inside him and get on his tail. We were at 500 feet at this time and he tried to break [veer] down. I followed him down to the treetops firing one gun. . . . The last I saw of the 109 was when one wing hit the trees and he did a cartwheel." The Messerschmitt was dead. Godfrey went on to damage an Fw 190 in the air and shoot up two other planes and two locomotives on the ground.

Lieutenant Archie W. Chatterley, also of the 4th, had a field day going through enemy airplanes like a flashing sickle. Chatterley's section spotted vapor trails; then six Fw 190s veered in from underneath to attack the American sky rams. "Immediately afterwards I saw a 110 attacking a Fort that had pulled out of formation. A few strikes . . . and when I closed [from the rear toward the back of the 110], many strikes were seen with pieces flying off and much smoke coming back over my windscreen." One victory for Chatterley. Almost at once he pounced on another Me 210 from behind, diving from 28,000 all the way down to 2,000 feet.

The Messerschmitt was "doing such turns, never straightening out of a turn until the very last. I was on the red line of my air speed when someone said he [the German] was shedding pieces. . . . He pulled out, made a zoom, dived and crashed." Another kill.

Chatterley shucked out the dead Messerschmitt and dashed on.

Next, he teamed up with three other Mustangs to attack an Fw 190.
"I asked Major Mills [on the radio] if we were going to queue up.
'Hell no, first one there gets him,' he answered."

Mills and Chatterley roared in with the speed Edgar Schmued
had designed into the P-51 and the power its Merlin engine had
developed. "When you were up there dogfighting or strafing," recalls Colonel Joseph A. Peterburs, another Mustang jockey, "there
was a noticeable difference. . . . It was years ahead in technology
over the P-40. You could do a lot more acrobatics."

Mills and Chatterley now closed on their quarry, loosing the
destructive force and versatility of the new species in the skies. "Major Mills edged the rest of us out and I can confirm his claim—after
seeing many strikes all over e/a [enemy aircraft] as it dove straight
in. Major Mills called on [radiotelephone], 'I got the pilot. I got the
pilot.' Lt. K. G. Smith and I heard this." Chatterley racked up one
more Me 110 destroyed, a shared Me 210 destroyed, a Ju 88 damaged on the ground and two more aircraft damaged on the deck, as
well as two locomotives, one of them shared with Godfrey.

Mustang power rolled out, for the first time, to smash the Nazis over their capital itself. All over the sky, they were battling
back the German defenders, cutting them down out of the clouds,
hitting the enemy hard. The bombers got in their share of the hunting, too.

Lieutenant William A. Kazlawski, navigator of a bomber in the
379th Bomb Group, recalled: "An Fw 190 came towards our plane
low at about 02:30 o'clock. When he was 600 yards away, I fired a
long burst at him. My bombardier reported that the Fw 190 blew
up and pieces went down in flames."

Tail gunner Sergeant Dominick R. Giordano, in another 379th
Bomb Group sky ram, saw "[an] Fw 190 was making persistent attacks at 0600 when we were straggling and by ourselves. E/A

[enemy aircraft] came in level at 0600. I let go 50 rounds up till 400 yards when he suddenly dove smoking and then broke into flames which completely enveloped him."

There now unfolded across an immense area 833 square miles in size, along a front of 15 miles, a spectacular tarantella of combat that filled the skies with fire, smoke and flak. From Erkner, 15 miles to the south of Berlin; to Potsdam, 13 miles west; to Oranienburg, 20 miles north of the capital, chaos and fury filled the chilly skies. German fighters with green, black and yellow markings swooped and sliced through the bomber formations, on the attack.

Captain Bernard L. McGrattan of the 4th, an ace with nine kills in all types of aircraft, dove on a German Ju 88 fleeing from a flock of B-17s after firing a salvo of rockets into them. McGrattan followed down through the altitude and fired at 300 yards. Pieces of the disintegrating 88 flew back, and the Junkers went pirouetting down, out of control. One Boche was finished. McGrattan climbed back up toward his bombers, spotting a speck he saw in the sky; the speck was firing at a distant chip. The chip was a B-17 Flying Fortress that burst into flame, falling out of the sky with 10 men aboard. Four chutes appeared. The rest of the crew went down with their bomber. McGrattan watched briefly, but there was no time: the vast action continued across the sky.

Blakeslee got another bandit, an Me 210. First Lieutenant Howard N. Moulton, Jr., was flying southwest of Berlin at 20,000 feet when an Me 110 came diving past him and a wingman. "As we went down after him, I followed about 500 yards behind Lt. Whalen. . . . He gave a short burst which was very accurate. The e/a blew up and many pieces flew back. . . . This was the last time I saw Lt. Whalen. . . . The e/a did a three-quarter spin and crashed." Whalen had downed the German. Moulton raced on, shooting ahead. After a tangle with another of his wingmen against an Me

110, which they destroyed, an Fw 190 shot past them down through the altitudes; Moulton was on him. "I did a turn left and then right, pulled out in back of him.

"The e/a started firing a second before I did. . . . I fired as soon as possible and saw strikes on both wing roots. The e/a broke into the clouds. I ran out of ammo after a burst of about ten rounds." Moulton's score for the day: one Me 110 destroyed (shared) and one Me 109 damaged.

Seconds flicked past, frozen in the clarity of battle; half-minute intervals dragged past in the tumult of combat straight across the ceiling. Action shook everywhere in the sky.

The cramped cockpit of a P-51B Mustang was the sleek, advanced chariot of a virtuoso warrior. It was compact, though long enough for legs to stretch, and warmed by the heat of the engine. Before the pilot were dials reading out compass heading, altitude, aspect of aircraft, oil pressure and temperature, speed and much more. The trigger firing the Mustang's guns was a red button on the top of the flight stick between the pilot's legs; when he squeezed the red button, four .50-caliber machine guns mounted in the wings shot at once, each spitting 13 rounds per second, 800 rounds a minute. The rest of the cockpit was a multicolor patchwork of gauges, switches and electrical rigs on both sides, alarms and toggles and, to the left, the throttle, fuel mix and propeller-pitch controls. These three handles had balls on their tops, hence the phrase "balls to the wall." Finally, there were pedals to control the rudder.

Battle was fast, taut and tense. In the heat of combat, across hundreds of square miles of ceiling and flocks of dozens of fighters, friendly and enemy, combat passed without letup or pause, in split seconds.

Here, at the very center of the mission, over the target of Berlin, the Mustangs were cutting their teeth.

In all, three entire fighter groups of P-51B Merlin Mustangs—the 4th Fighter Group, the 354th and the 357th, perhaps 195 Mustangs in all—had been assigned to the central objective of the raid, to cover and defend the invaluable bombers over their targeted drop zones, far inside the interior and beyond reach of other fighters. With limited range, the P-47s and P-38s, as over the past months, had been assigned to provide approach cover between England and the European coast and withdrawal cover back to England from the North Sea. They could go no farther. No Allied fighter had ever reached this point; no American bomber had ever penetrated 350 miles into Germany. Now the Mustangs—mixing with the Messerschmitts and Junkers, above, below, to port and starboard—were contending with the Luftwaffe.

Battle raged up and down, from 20,000 feet in the air to the turf four miles below, across a vast expanse of the enormous Berlin area, in the split seconds of plunge and maneuver that decided a dogfight. Mustang pilots reacted with lightning speed, almost by rote instinct. Colonel Joe Peterburs of the 20th Fighter Group recalled a fraught duel when an Fw 190 tried to ram him straight on: "We got hit and all hell broke loose, and the bombers were being decimated.

"I was looking out, and parachutes were down there, engines were falling, wings were falling and I was climbing out of the debris." Peterburs was more worried about being hit by the pieces of stricken bombers than by enemy fighters.

"And this 190 was coming right at me. I just turned in to him and started fighting. I could see, around his engine, he had 20-mm cannon, and I could see the little [gun muzzles] blinking like Christmas tree lights at me. I went back with my 50s and saw a couple of hits. . . . I passed under him about 50 feet. And as I passed under him, my flight leader just blew him out. . . . It was [a game of] chicken."

The sky was now a cauldron of smoke, action flashing among the bombers. The Fortresses and Liberators rammed on, closing steadily, unable to swerve, climb or drop, in order to maintain a level bombing approach.

Elsewhere, these scenes: "B-17 going down, one chute seen, 2nd chute got caught on Ball Turret."

"B-17 went down. Direct flak hit blew him off. No chutes seen."

Then they were over the target, bombs away—600 heavy Eighth Air Force ships raiding the German capital with 500-pound and 1,000-pound bombs. Each plane had 30 seconds over its designated target, dropping long-descending strings of bombs. The payload rained down on the city, flight after flight. Captain Andrew K. Dutch, a group navigator: "Bombs were away on Berlin at 1321 hours at 21,000 feet for the lead ship." They hit built-up areas in the east-central portion of Berlin with results "believed to be good." The big sky wagons passed on as if on a conveyor belt.

The unending line of bombers struck near Tempelhof Airdrome, in the southwestern suburbs of Berlin. Targets were plastered in the Spandau section, where the main railway was severed.

The Mustangs romped through the battlefield in the skies, unleashing the power and speed of Tommy Hitchcock's shrike and Edgar Schmued's vision. They performed with the versatility of a thoroughbred and the endurance of a quarter horse.

"It was sort of like part of your body. You think about turning left, and go left. You think about turning right [and bank right]. That's how it was with the P-51. I would think about a dive, it would do it. It just feels like a part of you. I'm sure it's the only aircraft I've ever felt that about," said Colonel Peterburs about Schmued's airplane.

First Lieutenant Alexander Rafalovich tore after a streaking Me 109: "I held my fire until I was sure it was not a P-51. Opened fire at 250 yards and blew his belly tank off," Rafalovich recorded. "My gun sight burned out. I pulled up and saw smoke coming from his engine," Rafalovich stated. "I then gave another short burst." Strikes again on the wing roots. Rafalovich went into a slow, sliding turn. "I gave him another burst and saw strikes. . . . He split-S'd [an evasive maneuver] and crash landed in the woods." A victory for Rafalovich.

All through the cold, late afternoon over the German capital, the Mustangs contended with the best German fighters, giving as good as they got.

The great aerial contest went on across the skies of greater Berlin. There were instances of phenomenal luck; there were instances of the worst luck. There were dodges closer than a matador's cape in a bullring; there were passes slicker than silk.

Captain Nicholas Megura, another 4th Fighter Group ace, spotted a flight of more than 20 single-engine German fighters covering a group of rocket-firing twin-engine Luftwaffe vipers. "I jumped three . . . Me 110s just as they let go their rockets, which burst behind the last bombers," Megura related in an after-battle report.

"I raked the 3 Me 110s which were flying wing-tip to wing-tip. As the #1 e/a broke into me, I saw strikes all over the cockpit and both engines as he disappeared." One down. "I cleared my tail. . . . I closed firing on the #2 Me 110 and saw strikes, and pieces falling off, and an explosion in the cockpit. I pulled up over him and last saw him in a vertical dive, pouring out black smoke." Two bags.

Megura raced ahead. Moments flew past, seconds ejected like bullet slugs.

"I climbed starboard toward a Me 110 who was climbing behind bombers. The e/a started violent evasive action toward the deck. There I closed him with my one working gun. . . . I saw strikes on port engine and cockpit. . . . Port wing tore off as e/a hit the ground and nosed over." Another kill. Megura next went after an Fw 190, but when he pressed the trigger on his guns—nothing. Nothing fired. "Hit deck, tried guns again on train with no results. Came home." The sun was sinking, the air colder, the smoke of battle beginning to settle.

Then it was out, up, away, back through the flak emplacements to the coast, across Hanover, Bremen, Amsterdam, flying northwest in the dying afternoon. They were not home free until they hit the Channel. The B-24 *Reddy Teddy* had been hit with antiaircraft and rocket fire, and was feathering on two engines, transferring gas to the working Pratt & Whitney R-1830-65 radials. The pilot, Captain Glenn E. Tedford, brought the *Reddy Teddy* through the last, tense miles of the way home.

In the last hour of the raid, even some survivors became targets. The B-17 *Vapor Trail* had gotten as far as the Zuider Zee in Holland with its crew of ten, when antiaircraft fire ripped through the plane.

"One shell passed through the wing in the 'star emblem area' and one exploded in the wing towards [the] tip," recalled the pilot, Captain Philip J. Field.

"It was the tail gunner who sprayed a criss-cross pattern on the anti-aircraft gun position. This gave us time enough at air speed of 105 mph to get almost out of range," Field remembered. "I don't recall any trouble from there on to base, other than to manhandle the controls."

Then they came in as the sun sank, winging to their bases, scattered like a constellation of dots across East Anglia. For some Mustangs it was a rough road home. It was wincingly cold four or five miles up; relief tubes for urination would freeze; some men just wet their pants, which froze. The pilots' feet were numb; their hands purple; some were bathed in fright sweat, but they were flush with the enormous satisfaction of hitting the "Big B." From the weary ranks of the worn but elated 4th Fighter Group jockeys, to the haggard bomber groups, they touched down vindicated. The triumphant 334th, 335th and 336th Fighter Squadrons of the 4th landed among the green fields around their base at Debden.

The bombers came in, too. Lieutenant Colonel Robert S. Kittel, who had watched bombers shot from the sky over Berlin, landed at 4:39 in the evening of that long, cold Monday; Lieutenant Cecil Roach's B-17 came in at 4:57; Lieutenant Victor J. Slack, pilot of another 379th Bomber Group B-17, touched down at 5:00.

The warriors came back in any and every condition. Captain Bernard L. McGrattan, who had been in the intense dogfight earlier that day, landed without brakes, trims or flaps; the ground crew got the crash truck ready for him. He skidded downfield and smashed into a hedgerow. The whole day he had been on the edge of survival. Now he was safe. The pilots, stiff and worn and barely mobile, were helped out of their Mustangs. Now the fighter jockeys went for debriefing, welcomed with a shot of whisky the intelligence officers had waiting on the table for them.

In the barracks, there were often harder tasks. Squadron execs went through the personal items of men killed in action, preparing them to be sent home. Intelligence was readied for teletypes to higher authority: the 4th Fighter Group had destroyed 17 planes.

The Battle of Berlin was the biggest air battle fought up to that time, with the most intense fighter offense the "Mighty Eighth" had ever put up. A force of some 660 Flying Fortresses and Liberators escorted by hundreds of fighters, led by Mustangs, dumped 1,626 tons of bombs on Berlin. Of those bombers that had carried out the strike, 69 had been shot down, about a ten percent loss rate, far improved over losses on unescorted raids the previous fall. In return, the Germans had lost at least 66 fighters and fighter-bombers; the buccaneering Mustangs and pursuit ships of the Eighth claimed an astounding eight-to-one kill ratio over Berlin that day.

The raid was a key morale success for the Allies. P-51s had decisively reached the capital itself.

Brigadier General Jesse Auton, commander of the 65th Fighter Wing, celebrated with the men of the 4th Fighter Group, issuing this bulletin:

> I desire to commend each one of you for your outstanding performance in combat over enemy occupied Continental Europe today.
>
> Your aggressiveness in seeking out and destroying the enemy definitely establishes the 4th Fighter Group as an outstanding fighter organization.

The thrill was felt by every man down the ranks who had flown in the raid. "On March 6, we finally made it to Berlin," wrote Lieutenant Gilpin, the navigator in the 381st Bomb Group Flying Fortress who had watched Mustangs in battle. "They said we could never do it. Most of our pilots didn't think we would ever get back, but we all went anyway." Gilpin and his men had proved something to themselves, and to the Germans.

The raid was a stunning blow to the German people—an open declaration that no corner of the Reich was safe anymore, even Berlin. The mission was a turning point in the war. Clearly, the Allies had both the range and the greatly improved fighting ability to reach and counter the invincible Luftwaffe—both capabilities directly attributable to the P-51B Mustang.

And the strike was a dark omen for the Nazi juggernaut. The massive assault of the raid stunned the Luftwaffe and Reichsmarschall Hermann Göring, chief of the German air force and one of Hitler's most trusted aides.

"The first time your bombers came over Hanover, escorted by fighters, I began to be worried," Göring later told an Allied interrogator at Nuremberg. "When they came in with fighter escort over Berlin, I knew the jig was up."

The P-51B Mustang, after a long odyssey from Edgar Schmued's drawing boards in Los Angeles to the front lines in the European skies, had dealt a body blow to the Germans from which they would never recover. The Mustang had broken the war wide open, as Hitchcock, Harker and the British RAF had believed it could; its vital strategic importance had been exactly as they had foreseen. All of Germany was an open target now.

CHAPTER 17

Death of a Jockey

One early morning before sunrise at the end of March 1944, Tommy Hitchcock stood on the tarmac at Debden, watching his nephew Avy Clark take off on a mission he would lead over Germany. The sleek, glinting Mustangs sped down the runway in groups of three and lifted into the sky. By now Avy was the number two on the 4th Fighter Group base, alternating with Don Blakeslee as commanding officer. Tommy had felt an almost paternal pride watching Avy in the briefing room earlier describing the day's mission to his pilots. And now, as he scanned the sky for Avy's Mustang at the head of the formation as it disappeared into the clouds, he thought about the 4th's recent record. In one month of flying P-51s, the group had shot down a whopping 160 enemy planes, compared to 120 over all the previous year, another astounding piece of evidence of the P-51B's firepower and capability.

By April air superiority over Germany was swinging back to the Allies, thanks in large part to P-51B Merlin Mustangs. Eighth Air

Force fighters were at last inserting bombers all across the Nazi empire, hitting Schweinfurt again, in addition to Frankfurt, Munich, Berlin, Mannheim and Ludwigshafen. Even Germany's staggering aircraft production could not fill the gap; there were still enough planes, but too few experienced pilots left alive to fly them.

With the Allies in charge of the skies, the sea war on the Atlantic tipping to the massive Allied fleets and the Russians breaking out of Poland toward Berlin, planning for the climactic Normandy landings could now move ahead.

The past months had taken Tommy far. In February 1943, a year before the P-51B began to prove itself in combat, he had returned to England, flying the embassy's Beechcraft all over the country to gather information on technical and training developments in the RAF and the Eighth Air Force. But he was restless. He wanted to be in the action flying. Although he was several years past forty, Tommy enrolled in RAF gunnery school, learning fighter-to-fighter tactics. He took a second RAF course for would-be group and wing commanders, flying in mock skirmishes over Gloucestershire with instructors playing the enemy. But it was not enough.

For months he had campaigned to lead his own fighter group— unlikely, given his age, even with his wide circle of friends in the Eighth. But suddenly, miraculously, the orders came through: in December he would be assigned to lead a fighter group in training stateside.

He arrived in Texas on Christmas Eve 1943 to take over the 408th Fighter Group, determined, now that his dream had come true, to make the most of it. In fact, the assignment delighted him. He led the group on practice runs and lectured on aerial combat. Peggy arrived, and they took a little house on the dusty outskirts of Abilene.

And then in early February the dream ended: the 408th was

broken up, the pilots ordered to England to replace others killed in action. Tommy returned to England, too, this time as deputy chief of staff of the Ninth Air Force Tactical Air Command, then forming up to provide aerial support for the Normandy invasion. Passing through New York en route to London, he had a few short days with Peggy, now back at home, and the children. When it was time for him to leave, his daughter Louise had a feeling of foreboding about his departure. "After he left for the airport," she recalled, "I went into the library. I looked at the indentation where he [had] sat. And it came into my mind that was the last time I would see him."

In early 1944, with D-Day looming across the sprawling map of German conquests, the P-51D model Mustang had appeared. This was the ultimate, definitive Mustang, the most numerous production run with 7,956 aircraft—the Mustang that would do the lion's share of defeating the dark condors of Hitler's air force. This most modern, most powerful third genotype of the original NA-73X had more firepower—two extra guns for six .50-caliber machine guns in the wings—and a speed of 437 miles per hour. The D was on the cusp of the jet age and looked it.

The breakthrough hull form, its modern streamlining and gunnery, made it look futuristic, and it was, with a bubble cockpit canopy for better rear vision and lines more suitable to a spaceship. The D fuselage had contours far more subtle than those of its predecessors, lean, elliptical lines and the familiar underslung air intake beneath the cockpit, which gave it a menacing look. The D concept was heavily armed and faster, and it had better striking power.

"It was a beautiful little plane. . . . It was gorgeous, a great looking plane," then-Captain Tom Stanback, assistant base adjutant of the 20th Fighter Group, says of the D today.

"It was a whole different plane, smooth flying; it responded quickly to any maneuver. It was fun to fly," says William R. MacClarence, then a lieutenant in the 339th Fighter Group, about the P-51D. "The transition . . . was going from a good airplane to an extremely good airplane."

The D would be a champion, an ace, but at its inception it had a serious flaw to surmount.

In the first few months of 1944, as Tommy was settling into his post at the Ninth Air Force, a bizarre and frightening pattern began to reveal itself. Suddenly some of the new-model P-51Ds were going into a steep dive that could not be controlled, and plunging to earth. The new fighters, according to General Pete Quesada, then head of Ninth Air Force Tactical Command, were "just diving into the ground. We couldn't understand it, and Tommy couldn't. Obviously you can't have a useful force that is going to destroy itself."

Engineers suspected a new gas tank in the P-51D—separate from the internal 184-gallon tanks and 108-gallon drop tanks already developed—was at fault. The additional reserve tank, behind the pilot and the cockpit armor, could be destabilizing the plane and sending it into a nosedive. The new addition held 70 gallons; it was important to the long, extended range that was the D Mustang's unique capability. Could this extra tank be causing the crashes?

It was a dark time. In March 1944, Tommy's nephew Thomas Hitchcock Clark was shot down and killed in a Spitfire. Now, in April, the P-51D of Peter Lehman mysteriously dove into the ground during a raid. Peter was the son of Herbert Lehman, the former governor of New York and founder of Tommy's firm on Wall Street. The big, good-natured father of two had flown for the RCAF before the U.S. entered the war; now he was a fellow pilot of Avy's in the 4th. His death in a plane Tommy had championed was intolerable.

"It was perhaps our most serious worry that spring," said General Quesada. "We found that we were getting an extraordinary number of airplanes—apparently without pilot error—just diving into the ground." The problem must be solved for the bomber missions to continue. There was no time to waste.

Tommy, in charge of research and development, had overall responsibility for investigating the mysterious accidents, but he did not have to test-fly the P-51D himself. There were pilots on his staff whose job it was to do it. It was not his assignment. But with typically decisive boldness, Tommy set out to solve the problem on his own.

On the morning of April 18 at about 11:00, when he had finished all the paperwork in his office, he drove out to an air base near London. A P-51D was waiting for him, fueled and armed. Tommy climbed into the cockpit, buckled his harness, fired the engine and took off. He climbed into the sky and headed for a bombing range not far away. An English springtime haze covered the ground below; he soared away from the fields, speeding through the ceiling over Hampshire, veering, climbing. Reaching the target area, he put the plane into a steep nosedive. He could not get it out. Spiraling and spinning through the altitudes, wheeling and turning down through the sky, the plane crashed into the ground. A curl of black smoke billowed up from the earth. Hitchcock's body was found near the wreckage. The Mustang's most important apostle was dead. He had died rushing directly into a challenge with the courage that had always been a hallmark of his character.

Tommy's daughter Louise, 14 years old at the time, remembers the moment she heard of his death. "I was by myself in my mother's apartment in Gracie Square," she remembers today. "And I was washing my hair and the phone rang and it was the *New York Times*, and they said, 'I'm just confirming the death of Col. Thomas Hitchcock.' And I said, 'No, you're wrong, it's Thomas Hitchcock Clark,' because he'd just been killed the week before. I thought it was a mistake. Then Mr. Charlie Schwartz, who was a great friend of my father's, rushed in and threw his arms around me." Louise's premonition had come true. She would never see her father again.

Tommy's stepson, Alex Laughlin, then a 19-year-old radio operator in the Air Transport Command in Gulfport, Mississippi, recalls his death: "My mother was devastated. She had lost two husbands. She was only forty-four years old . . . with a family and four [other] children." At his age and with so many reasons to live, many men would not have taken the risk. But that was the kind of man Tommy was. "He was a bulldog," remembers Laughlin, "about something he really believed in." His determination overcame any possibility of fear.

Tommy's last letter to Peggy, written the day before the crash, was filled with concern and longing for her and their children. "Don't forget," he signed off, "that I love you." At the same time, his pride in the Mustang was clear. "[Avy's] group," he wrote, "has been going great guns since they went over to Mustangs. They are now the highest scoring U.S. group in England. . . . They are making deep penetrations into Germany and chasing German planes all around the treetops."

Back in Inglewood, California, Edgar Schmued soon isolated the problem with the brand-new Ds. The failing lay in the ammunition boxes for the fighter's .50-caliber machine guns installed in the wings. Schmued and his engineers determined that the covers for the ammunition boxes, despite being made of sturdy one-eighth-inch aluminum alloy, tended to buckle and bend, changing the curvature of the wings and suddenly increasing lift so forcefully that the wings ripped off. The extra fuel tank behind the cockpit had not been the problem at all. The defect was corrected, and there were no more problems with the P-51Ds.

Schmued was greatly saddened by Hitchcock's death; the two had become friends in an eight-week visit Schmued had made to

England during 1943. That spring, Tommy had helped the Mustang master in a quest to build a fighter even faster than the P-51D. In February, Kindelberger had sent his designer to Britain—his high-priority assignment to confer with British aircraft manufacturers and RAF officials about weight and engines for new Mustang models already on the drawing board. From the ambassador's residence in London, Tommy had cut through wartime red tape to arrange priority transport for the trip.

The Mustang inventor waited eight days in New York for the ice to clear, then boarded a Boeing 314 Clipper flying boat, a huge slow-flying seaplane like the ones Tommy had invested in at Lehman Brothers that would take him across the winter North Atlantic to England. There, the ever-generous lieutenant colonel would become host, guide and advisor for Schmued's fact-finding tour.

Hitchcock and Schmued journeyed to the de Havilland plant to inspect an early jet. Schmued, intent on new Mustang types, then went to Rolls-Royce to look at their latest version of the Merlin—the lighter, more powerful RM.14.SM. "This was the answer to our many prayers," recalled Schmued, "more power without increasing weight. Every airplane designer likes to hear that and so did we."

Weight had become a concern of the Army Air Forces. "American airplanes were heavier than the British ones, and they wanted to know why," remembered Schmued. Schmued sent his team around to the British aircraft manufacturers to weigh the components of all the British fighters they could find, especially the Spitfire. "We used all these load factors to help whittle out a good deal of weight in a new design, which we called the P-51F and G."

"I went home, after being flown around by Colonel Hitchcock, who took care of me completely. Wherever I wanted to go, he flew me with his private plane, a reverse staggered Beech biplane," Schmued recalled. The duo sojourned together from Derbyshire to

London, Schmued walking carelessly through the blacked-out city streets as shrapnel from antiaircraft guns rained down on him. The American blue blood and the German émigré, who had never met before, became fast friends. The trip would lead to the development of the F, G and H models of the P-51—700 pounds lighter and 60 miles an hour faster than the P-51D. As a result of this trip, Hitchcock's legacy to the P-51 would extend even further into the future, years beyond a life cut short.

Hitchcock was killed just six weeks before D-Day, the invasion for which his fighter plane had broken open the way. The *New York Times* ran Tommy's obituary on the first page. Arthur Krock wrote: "His sense of duty toward every obligation of life . . . was permanent in his nature and it was this that brought him to his death." Hitchcock had died, Krock said, "when there was great need for him to live."

In a separate tribute, Arthur Daley, the great *Times* sportswriter, recalled Tommy the polo star: "The man had no fear. He had showed that on the polo field. Tommy drove through with that phenomenal concentration of purpose which dominated him completely. Nearly every spill or collision he had was due to the fact that someone neglected to give ground to him. He never gave ground to anyone."

Tommy Hitchcock was a visionary who had not only understood the phenomenal promise of the spurned-stepchild Mustang fighter plane but had also fought its cause to the top of the air force until its role in the faltering air war was certain. He had led the drive for mass production of the Merlin Mustang, knowing that Allied triumph in the sky war depended on it. His life had impelled the greatest fighter plane to come out of World War II. But he did not live to see the final sweep of victory his efforts had brought to history.

CHAPTER 18

An Endless Roar Overhead

For days the weather had been socked in, gray, forbidding, and the commander of the invasion, Major General Dwight D. Eisenhower, had hesitated, held back. The navies, armies and air forces of America, Britain and Canada, which had practiced the assault for months, were harnessed and ready for the massive operation that would comprise the equivalent of 15 army divisions. Command of the skies over the Normandy beaches would be key on invasion day. All across the nebula of airfields in southern England, pilots and crews waited.

Weather was critical. Group Captain James Stagg, Ike's Scottish chief meteorologist, had been briefing him and the generals every 12 hours about the forecast. His predictions had already led Ike to postpone the invasion from June 5 to June 6. At 4:00 a.m. on the morning of the fourth, as the wind whistled and rain pelted the windows and Eisenhower paced up and down in the war room at Southwick House, Stagg announced a break in the weather—a clear spell beginning the night of June 5. Ike continued to pace back and forth in the haze of cigarette smoke. His open, warm manner and usual wide Midwestern grin were absent now. A chain-smoker, he was exhausted by the magnitude of the decision he alone must make as supreme commander of the Allied Expeditionary Force. After five minutes of silently treading the floor, he spoke: "OK," he said. "We'll go."

At 4:15 a.m. the order went out. Overlord was on.

That evening Churchill and his wife dined together in the bomb-proof underground Annexe in London and then made their way to the map room to view the disposition of airborne troops set to deploy into France. When Churchill retired for the night, the first were parachuting into the Normandy countryside. As Allied forces made their way across the Channel, FDR announced in a radio address that evening in Washington that the Allies had captured Rome, knowing as he spoke that Overlord had begun. Before Eleanor went to bed, he told her the attack in Normandy was underway. With four sons in the war, she felt great anxiety: "I feel as though a sword were hanging over my head," she wrote, "dreading its fall and yet know it must fall to end the war."

At Debden, too, the 4th Fighter Group and its commanding officer, Don Blakeslee, were as tense as piano wire. Several times Blakeslee and the 14 other fighter group commanders in the Eighth Air Force had been "summoned to headquarters for invasion briefings, only to have them scrubbed because of bad weather." But late on the afternoon of June 5, Blakeslee could be seen running into the officers' mess with a large sealed manila envelope under his arm. Ground crews could be seen placing 1,000-pound bombs beside each plane and painting "large black and white stripes on the wings and fuselages of the Mustangs" so they would not be mistaken by friendly fire. His pilots suspected the landings were on.

Blakeslee briefed them at midnight; by then airborne divisions were already flying across the Channel to lead the assault as the first boots on the ground. Blakeslee was grim and blunt. The Germans were rumored to have 2,000 aircraft ready to protect the Atlantic wall. "I am prepared," he told his men, "to lose the whole group."

At 3:00 a.m. the pilots of the 4th walked to their planes and were strapped in. Some 60 Mustangs started their engines and joined the flight line. They taxied away in the darkness and, one by one, took off behind their leader. The mechanics could see their red and green navigational lights moving across the field as they formed up. Then they took off into solid clouds, heading south toward the Channel.

All that night bombers and fighters collected in the skies over East Anglia, an immense convoy of some 3,000 aircraft and 900 gliders making an endless roar overhead. All who witnessed it would say the deafening rumble of the planes and the sight of the wings and the combat boxes was the most awesome thing they had ever heard or seen. Airplanes, ships and the men in their landing craft all prepared themselves for touchdown on the shores of occupied Europe, across the 30 darkened miles of the English Channel.

Then, at 6:30 a.m. on June 6, 1944, the greatest amphibious invasion in the history of warfare stormed onto the Normandy beaches of France. Three thousand landing craft, covered by a flotilla of 500 Navy ships, made their way onto the American beaches code-named Omaha and Utah; the Canadian beach, Gold; and the two British beaches, Sword and Juno. More than 156,000 troops—about ten army divisions—began to wade ashore amid a fusillade of gunfire.

By 9:00 a.m., coastal defenses had been generally breached; a beachhead was established not long after. On Sword, Juno and Utah beaches, resistance had been light. But American soldiers at Omaha Beach and the Canadians on Gold Beach ran into furious opposition. Oncoming waves of men were met by intense machine-gun fire from the sheer, steep cliffs overhead. The German batteries on the bluffs—hooded black gun emplacements that stand today—poured fire into the array of arriving forces and fleets. Two thousand American soldiers were lost in the first hours of the invasion; the first Canadian echelons ashore suffered a 50 percent casualty rate. Total casualties for D-Day were estimated at 4,400 troops, with thousands more wounded and missing. Today, in the gentle summer sun, the neat, pristine crosses over the graves and the stars for Jewish soldiers still stand in a magisterial quiet in the cemeteries stretching away above the beaches where the invasion came ashore.

But casualties could have been far higher.

The British had been especially pessimistic. Before he left Clementine for the night on the fifth, Churchill had said to her: "Do you realize that by the time you wake up in the morning twenty thousand men may have been killed?" At a briefing earlier he had shaken his head: "I wake up at night and see the Channel floating with bodies of the cream of our youth." Air Chief Marshal Sir Trafford Leigh-Mallory had predicted the airborne attacks preceding the landings would be "futile slaughter," and the landings themselves a failure. And after giving the order that set the invasion in motion, Eisenhower had scrawled a statement to be released in the event the invasion failed. "If any blame or fault attaches to the attempt," he wrote, "it is mine alone."

On D-Day, Eighth Air Force fighters were to form an airborne wall to the south, east and west to prevent the Luftwaffe from reaching the landing area. Blakeslee's 4th flew nonstop between 3:00 a.m. and midnight on the sixth, returning to base only to refuel and rearm. The Mustangs mounted bombs and delivered them at 12:20 p.m. and again at 6:20, discharging their payloads on German troop trains and bridge crossings, keeping the Wehrmacht at bay while

Allied soldiers, their ordnance and their equipment were moved ashore.

Major James Goodson of the 4th, an ace with 15 aerial victories who flew on D-Day, recalled: "We had told Eisenhower and his staff that we could prevent the German air force and army from moving up to the beachhead during the hours of daylight . . . and we spent the day proving it. We attacked everything that moved on the roads and rails of Normandy leading to the landing area. The destruction was terrible. We strafed columns of trucks and light tanks, while the Germans leaped from their vehicles and dove into what cover they could find at the side of the roads, or desperately tried to set up flak guns before we could mow them down. We sprayed the columns of marching men and watched them fall as our six machine guns mowed them down."

The 4th lost six pilots that day to enemy fire. Blakeslee and the group returned after midnight. As he taxied down the runway the landing gear of his plane collapsed, as if from sheer exhaustion. Blakeslee climbed out of his plane red-eyed and numb and gathered his men around him in the officers' bar. They drank a deep draft.

A German officer interviewed after the war who had been on the ground in Normandy was in disbelief about the performance of the 4th's Mustangs that day. "We couldn't believe the ferocity of the attacks, nor the tenacity of the pilots. They ignored the flak completely, and came in time and time again, almost touching the ground. . . . We never expected you Americans to fly like that."

The German aircraft Blakeslee had been told to expect did not materialize. Only about 200 were left to respond to the invasion. Their absence, on a day when the Allies put 11,590 planes, including 3,700 fighters, in the air showed the overwhelming superiority that Schmued's laboratory XP-51B had achieved in a few short months. The Luftwaffe, brought to its knees, could muster only the odd, isolated attack with little effect. Blakeslee and his men shot down only three enemy planes on D-Day. As American, British and Canadian troops headed ashore onto the Normandy beaches under heavy fire that morning from gun emplacements on the cliffs above, only two German planes showed up to strafe them from the air.

The epochal landings had been made possible by destroying Luftwaffe aircraft in the air and parked at airfields on the ground. And the strafing attacks of the Eighth's Mustangs, which strafed everything that "moved on the roads and rails" for weeks before the invasion, and on D-Day itself, blocked German reserves from moving up to the beaches from the rear as Allied troops were coming ashore. German officers would later insist this inability to move men and supplies to the front had scuttled their defense. They believed to the end that if reinforcements had been able to move toward the coast, the Allied tide could have been stopped. The Germans were stymied, forced into long detours and small straggling groups. No massed force could be gathered. Fanatically holding on to their defense, the Wehrmacht would fight on desperately for one more year. But their defeat had begun in June.

By dint of will and wits, the Allies built two entire artificial harbors on the beaches, Mulberry A and Mulberry B, to handle the tide of troops and stores coming ashore after D-Day. Over the next days they would unload 2.5 million men, 500,000 vehicles, including 14,000 tanks, and four million tons of supplies across their docks. The fight for France had begun.

It began in adversity, with woeful setbacks slowing any advance and the land armies bogged down in arduous terrain.

Summer found the Allies, more than a million of whom had come ashore just weeks before, bottled up and caught in the northward coastal jut of the Cotentin Peninsula, where the D-Day landings had unfolded. The coalition armies were ensnared in the difficult-to-navigate "bocage" or hedgerows, small fields surrounded by "six foot high earthen walls topped with trees and bushes." From Cherbourg toward the Atlantic, to Le Havre to the east, Eisenhower's divisions battered away in futility, trying to break out of Normandy.

In the muddy impasse, Allied troops found themselves stumbling through the bocage, advancing by mid-July only 25 to 30 miles—along a slender front barely 80 miles wide. One British armored thrust was defeated on June 13; a massive American infantry offensive failed on June 29. Total casualties on both sides amounted to 195,000 troops. It appeared the deadlock in the French interior was sliding back to the abyss of trench warfare of the First World War. General Omar N. Bradley, the U.S. First Army commander, had flown from France to England to beg for a massive aerial barrage to blow a hole in the German lines. Normally modest and retiring, the rough-hewn Bradley was now confounded and furious. "I want it to be the biggest thing in the world," Bradley had told General Eisenhower. The ensuing lightning bolt was called Operation Cobra.

On Monday, July 24, the Eighth and Ninth Air Forces set out to break open Nazi positions in France. Bad weather over Normandy caused a cancellation of operations, though not in time to stop the first waves of bombers from dumping tons of bombs on the enemy, including dozens of which landed behind American lines and killed 27 U.S. soldiers, wounding 131 more. The next day Cobra was uncoiled in all its massive aerial force. One thousand five hundred B-24 Liberators and B-17 Flying Fortresses accompanied by Mustangs dropped 3,437 tons of bombs on French soil; 559 fighter-bombers laid 212 tons of bombs with quantities of napalm over German units. The Mustangs attacked in both strafing and bombing runs.

It was the first major use of carpet bombing by the Americans.

Because the bombs were dropped perpendicular to the front lines instead of parallel as Bradley had ordered, again, tragically, 111 American soldiers were killed and 500 more were wounded in the confusion and chaos of action. Among those lost was Lieutenant General Lesley J. McNair, commander of the Army ground forces and the highest-ranking U.S. officer to be killed in the war, who was observing that day on the front lines. The stunning raid, pounding German lines, reminded celebrated war correspondent Ernie Pyle of "a storm, or a machine, or any resolve of man that had about it the aura of such a ghastly relentlessness."

The bomber crews took it as another job. To Lieutenant Robert G. Littlejohn, the lead bombardier for his formation in the 379th Bomb Group, it was a matter of cool exactitude. "Bomb run was 2 minutes in length after sighting target area through smoke and haze." Littlejohn's commander, the pilot of their B-17, Captain Rowland S. Williams, recorded: "We bombed the primary visually from 12,000 feet with pretty good results observed. . . . Area support by friendly fighters was very good." The divisions of Allied airpower, led by Mustangs, dumped stick after stick of ordnance into the narrow target area.

They bombed Montreuil; they bombed Saint-Lô. Enemy flak was "very meager, inaccurate" at the target; crews reported seeing columns of trucks and 30 to 35 small boats in the harbor at Granville. The raid loosed 3,400 tons of explosives onto a target roughly 1.5 miles wide, the pilots delivering this obliteration in the space of about an hour. The area, obscured by smoke and dust, was devastated. General Eisenhower, who had crossed the Channel to see for himself the situation in Normandy, drove down to Bradley's command post. He passed not far from where General McNair had died. Every village he saw was decimated. There was no sign of human life. He returned to England in the evening despondent over the losses to friendly fire, and uncertain about the results.

But in fact Cobra had done what it set out to do and more. German soldiers were shocked by the devastation. In two days their positions had been turned into a haunting no-man's-land by the pointblank attack. Communication with units to the rear had been completely severed, increasing their isolation. Wehrmacht major general Fritz Bayerlein, commander of the Panzer Lehr Division and a veteran of stunning defeats in North Africa, called the battlefield after Cobra a *"Mondlandschaft"*—a lunar landscape. He recalled the day in his memoirs: "The bombers came in as if on a conveyer belt. Back and forth the carpets were laid, artillery positions were wiped out, tanks overturned and buried, infantry positions flattened, and all roads and tracks destroyed. . . . The shock effect on the troops was indescribable. Several of my men went mad and rushed around in the open until they were cut down by splinters. Simultaneously with the storm from the air, innumerable guns of American artillery pounded drumfire into our positions. Over 70 percent of my soldiers [roughly 1,000] were either dead, wounded, crazed or dazed." All headquarters records were lost.

Operation Cobra opened the way for an onrush of stampeding Allied infantry divisions. On July 25, the American VII Corps—commanded by Major General J. Lawton "Lightning Joe" Collins, the scrappy little West Pointer who had grown up poor and barefoot in rural Louisiana—raced through the breach the Eighth and Ninth Air Forces had opened up. The following day, the German Seventh Army counted seven ruptures in its front. On the twenty-seventh, the 30th Infantry Division reported: "This thing has busted wide open." General Omar Bradley wrote to General Eisenhower on the twenty-eighth: "Things on our front look really good." The Allied air forces, bombers and fighters had provided superlative airborne artillery. As 100,000 troops poured through the hole in the German lines that Cobra had created, the Americans—now reinforced by the

197

flamboyant General George S. Patton, hell-for-leather tank commander of North Africa fame, and his Third Army—captured key bridges at Avranches, the gateway from Normandy into Brittany. American tanks and infantry flooded into the open country beyond. The Allied thrust to Germany was unleashed.

The eighty-day Battle of Normandy ended with the collapse of a desperate attack by Hitler's Fifth and Seventh Armies. Under fierce resistance from hundreds of U.S. and British fighter-bombers, the Germans retreated into the gap between Allied forces at Falaise. There, eight P-51s of the 354th Fighter Group took on seventy Fw 190s attempting to cover the German retreat and dispersed them. German tanks and troops trapped in the Falaise-Argentan pocket became sitting ducks for Allied aircraft, so thick overhead "that formations had to wait in line for their turn to dive in and attack." In the end, the survivors fled on foot through the gap, abandoning their arms.

Moving east on August 17, the Americans took Orléans on the Loire; on August 19 an American spearhead crossed the Seine River. A coordinated drive could be launched at last. Then history, for one burnished moment, glowed like the sun.

On August 25—called by some "the happiest day in the history of the world"—Paris was liberated. The underground French resistance rose up in the streets of the City of Light from the Arc de Triomphe to the Place Pigalle; that Friday a French armored division, led by General Jacques Leclerc, triumphantly entered the French capital and took the German surrender. Days of riotous celebration broke out in every quarter fanning away from the banks of the Seine River, capped by the liberation of the stately Ritz Hotel by the novelist Ernest Hemingway and his band of irregulars. At the hotel bar, the Ritz captain uncorked vintage champagne to celebrate. The day had relit the flame of the immortal city.

CHAPTER 19

Big Brother, Little Friend

The fighter war was now a three-front war: bomber escort, ground attack and reconnaissance. The bombers, "Big Brothers," were still the kingpins; they carried the bombs that would obliterate an objective. But they depended on their "Little Friends" the fighters to protect and insert them to the targets. The year 1943 had shown that no bomber payload was worth the explosives it packed without fighter escort.

Now the VIII Fighter Command had the Mustang, a doomsday weapon that could strike down any foe, outperform any attacker. The 14 newly equipped fighter groups that composed it were all splayed out in a fan from London, scattered across East Anglia from the Thames River mouth in the east; to Northampton, the capital of English shoemaking, in the north; to Oxford with its great university in the west. Bomber bases were intermixed with the fighter bases; reconnaissance groups were intermingled, too, across the same expanse of the southern midlands.

By the fall of 1943, VIII Fighter Command was led by Major General William E. Kepner, who had reluctantly yielded to Blakeslee's plea for Mustangs. He had served previously in the Marines as a cavalry officer, then as an infantryman in the First World War. Kepner was a pioneer balloonist and airship pilot whose remarkable career in the military began when he ran away from home in Kokomo, Indiana, to join the Marines. He personally took part in 24 combat missions in fighters and bombers during World War II. General Kepner's aerial swarm included everything: Lightnings, Thunderbolts and finally Mustangs; the groups each had distinct identities and each boasted its share of aces. The 4th ("Fourth But First") was led by Don Blakeslee and became known as the "Blakesleewaffe"; it had the highest number of kills in World War II. The 352nd had in its ranks the all-time top Mustang ace, Major George E. Preddy, Jr., an intense, skinny, dark-haired North Carolina boy who bagged a total of 26.5 enemy kills and once shot down six German planes in a single day. These were the bandits who were shattering Hitler's Luftwaffe, breaking open the whole air war.

As important as the sky-bound pilots were the ground bases below. The fighter base was home plate. Every fighter jockey flying in attack at the Luftwaffe above was supported on the ground by a village of some 2,000 to 3,000 crew and support staff, who ranged from intelligence and weather officers, to mail clerks, paymasters, cooks, armorers and mechanics, to adjutants, statistical officers and electricians. This variegated group of ground crews, clerks and senior officers was the launching pad for the Mustangs and the logistical wings of the fighters—as important as their real aluminum wings.

The fighter base was the hearth—tin Nissen huts or sometimes more comfortable brick buildings—to which the pilots returned to rest and sleep. It was a complete community, from chaplain to officers' bar, in which the men lived out their personal lives in the tumultuous time of war. The base and ground crew sustained and reinforced the machines and the men who did the fighting overhead. The base personnel were the arm that hurled the spear.

The busy teams of a typical base were as numerous and variegated as the population of a country town. As good an example as any such an encampment would be the 20th Fighter Group in King's Cliffe, Northamptonshire, which traded in its P-38s for P-51 Mustangs in July 1944. They had been among the raw recruits waiting to go to war in the hot summer of 1943.

That August, three trains had left March Field, California, with the men of the 20th on board. For days, as the trains crossed the country, the troops and airmen sat covered with coal dust and soot from the old stoves in each car. Arriving on the East Coast, they boarded the passenger liner *Queen Elizabeth* in New York Harbor. The great ship, packed to the gunwales with 19,000 men bedded down in double- and triple-decked bunks, traveled faster than Hitler's lurking U-boats, safely reaching Scotland in five days. There the men disembarked and scattered to bases all over southeastern England. At King's Cliffe, the men of the 20th found an old RAF base with dilapidated buildings mired in mud. But with energy and ingenuity, they fashioned a comfortable if spartan new home for the group.

The 20th Fighter Group base was a residential colony, a maintenance facility, a combat control center and an administrative arm that managed the air war fought by its squadrons as well as the daily routines and needs of its pilots and staff. Within these quarters played out the yearnings, ambitions, tragedies, hopes and victories of life during those heady years when the men were part of the 20th.

The barracks, all in Nissen huts, were heated by coal stoves; once every week or ten days, you got a shower. The mess halls for officers and enlisted men served their meals; the officers' club, decorated with a life-size pinup of a naked buxom babe, served hard liquor that you bought with your pay.

The aviation facilities themselves were the center of the base: the airstrip and the planes it sent into the air were the central crossroads of the camp. The field had three runways, one immense hangar for maintenance and several large areas where the planes were parked out in the open.

Other specialized groups scurried about, completing the blizzard of functions that kept the 20th going. There was an intelligence section, which studied enemy forces, action reports of pilots and data on targets and plans for slated raids. There were a weather office and a medical dispensary. The 20th Fighter Group was composed of three squadrons: the 55th, the 77th and the 79th—some 20 planes to a squadron, 3 squadrons to a fighter group, about 60 planes. The 55th had armaments sections, a communications section, a photo section and a refueling section, staffed by Corporals Philip Sharkey and Frank Agostinucci and Sergeant Harley Freese. The 446th Air Service Group of the 20th, under the command of Lieutenant Colonel Norvil T. Hinds of Caldwell, Idaho, was a byzantine warren of responsibilities, with a chemical warfare section, a firefighting section, a medical section and a utilities section.

And it went on and on, the nebula of activities that made up the base. An ordnance section dealt with arms and ammunition, with pyrotechnics (explosives) and spare parts for arms and vehicles. There were a finance office, an accounting section, a vault and a post office, plus a welding shop, a parachute shop and an electrical shop; the 1029th Signal Company provided telephone linemen and landing-light maintenance.

In the routine activity within this hub, the men slept, washed, ate, dreamed, shot the breeze, read, played chess and made fast friendships in lives now seized by war and rooted in tiny King's Cliffe, England.

"You talk about a family; in a sense it was," recalled then-Captain Thomas M. Stanback, Jr., an assistant adjutant with the 20th Fighter Group. "We all ate together, and sat around and played cards together." There were dances and poker games, endless bull sessions. For more diversion, airmen could go into the village of King's Cliffe, or, more enticing yet, the larger nearby town of Peterborough, where there were a movie theater, pubs, bars and shops.

"There were nightly passes," Stanback recalled. "The guys mostly headed toward the bars. If you went into a bar, the gals and men were together. I know the pilots had girlfriends, British girlfriends. The bars were the popular places," Stanback remembered. "It was a typical scene, throwing darts, the beer not being cold enough to suit the Yanks." Relations with the British could be edgy: a common British label slapped on the Yanks was "overpaid, oversexed and over here." Generally, though, the two populations got along.

There were social functions at the base, parties and games to which the local people were invited; refreshments were popular with war-rationed villagers. Friendships were struck up, some of them lasting. Some families adopted lonely airmen; some airmen adopted lonely children and widows; many romances flourished.

And, of course, there were many dark times when a pilot did not return from a mission, reported shot down or missing in action. "The word would be that someone got shot [down]," recalled Stanback. "Of course I didn't know many of the pilots, but I knew some of them, and you always remembered . . . when somebody didn't come back." These were the casualties who left an empty bed in the

Nissen huts where they had bunked: the casualties in what one young officer had called "this dangerous game": Second Lieutenant Edward A. Loetscher, from Ellington, Connecticut, killed in action on November 9, 1944; First Lieutenant Lauren J. "Mad Dog" Taylor, of Boone, Iowa, the holder of an Air Medal with three Clusters, with 95 hours and 55 minutes of combat hours, killed in a training accident on March 25, 1944; First Lieutenant Paul "P. D." Denbo, of South Gate, California, with 184 hours, 10 minutes of combat hours, an Air Medal with three Clusters and a Distinguished Flying Cross, killed in action on August 12, 1944. They were among the 89 men of the 20th Fighter Group who never came back, who perished in the fight to defeat Nazi Germany, giving their lives battling Hitler's Luftwaffe. They are remembered today by their brothers in the 20th who survived.

Over all these facilities and men—shops, sheds, huts, planes, duties, missions—was the group commander, usually a highly decorated colonel. The group commander often led the group into battle, flying at the head of the formation, and managed the entire operation on the ground. Colonel Harold J. Rau, of Hempstead, Long Island—with 275 combat hours, an Air Medal with eight Clusters, a Distinguished Flying Cross with two Clusters, a Silver Star and five kills to his record—was a popular group commander. Rau was "kind of laid-back, a good old boy," says Stanback. "He loved to talk." Rau would famously lead the 20th on its turn in the bombing shuttle to Russia, Operation Frantic, and report that the mission had been a great success—he had managed to see his wife, also in the military, on leave in Italy.

Colonel Robert P. "Monty" Montgomery, another group commander, was celebrated for having been shot down on February 11, 1944, evading capture in Spain and making his way back through enemy territory to King's Cliffe. Montgomery had 6¹/₃ kills to his record, four enemy aircraft damaged; he held the Air Medal and eight Clusters, the Purple Heart, the Silver Star, the Distinguished Service Cross (the second-highest American military decoration after the Congressional Medal of Honor) and the French Croix de Guerre avec Palme, among other medals. "He was a delightful guy. Big, good-looking," says Stanback. "He was very friendly and knew everybody's name and spent a lot of time just talking." Once someone had famously asked him how, as a big man, he liked the snug single-seater Mustangs. Replied the rangy Montgomery: "I just love to fly little planes."

Squadron commanders led squadrons: Major Richard L. Ott, who led the 79th Fighter Squadron, was from New York City and had 3.5 aerial victories, an Air Medal with four Clusters and a Distinguished Flying Cross. He was killed in action on January 30, 1944. Captain Jack M. "Happy Jack" Ilfrey, another decorated ace with eight victories on his belt, also commanded the 79th later on.

Beneath these, but as intricately tied up in the lives and times of pilots and crews, were the chaplain and the flight surgeon. Base chaplain Arnold E. Heimsoth, a Protestant, had two Catholic and two Jewish chaplains on his staff. Between September 1943 and May 1945, Reverend Heimsoth ministered to a total of 25,266 worshippers and conducted 583 religious functions. These included Sunday and weekday religious services, hospital and sickbed ministrations, Bible study sessions and prayers with the pilots before they took off on missions.

The victories of the 20th Fighter Group were won by all these men on the base, some 2,000 staff and crew, who fulfilled their many assignments and supported the wings first of the P-38s and then of the agile Mustangs sent by the 20th over Nazi Germany.

There were other famed groups. The 352nd, "The Blue Nosed Bastards of Bodney," was home to all-time top Mustang ace George E.

Preddy, Jr. The hallowed 4th displayed remarkable bravura from its commander down through its ranks. The unit bagged fully 1,016 planes destroyed throughout the war. They had a contest with the 56th Fighter Group (still in P-47s) to see who would score the highest number of aerial victories in battle: the 56th racked up 1,006.5 to the 4th's 1,016.

And there was yet another uniquely illustrious group—the 332nd Fighter Group, the segregated, all-Black unit that became one of the most successful and highly decorated fighter groups of the war. The legendary "Tuskegee Airmen," who flew in the Mediterranean with the Fifteenth Air Force, were so good that bomber pilots constantly asked for them as escorts. The 332nd was formed in 1943 under the command of Colonel Benjamin O. Davis, Jr., and it was composed of the 100th, 301st and 302nd Fighter Squadrons; it saw action over Sicily, Monte Cassino, Ramitelli, Berlin, Normandy, the Po Valley and many other targets. In dive-bombing attacks, bomber escort and ground attack, the Black pilots of the 332nd proved themselves to be virtuoso warriors, winning three Distinguished Unit Citations, 96 Distinguished Flying Crosses, 14 Bronze Stars and 744 Air Medals.

The Airmen came about in this way: "In the early 1940s, the U.S. Armed Forces were strictly segregated," according to Tuskegee Airmen archival sources. "Black men could serve in separate divisions or be assigned to support roles like cook, kitchen patrol, or grave-digging duty. Blacks were prohibited from being pilots; the military command thought they lacked the intelligence and reactions that flying demanded."

When America entered World War II, the Civilian Pilot Training Program was opened to Blacks who had finished two years of college; in 1941 the Tuskegee Institute was awarded a U.S. Army

Air Corps contract to train Black military aviators at a nearby airfield. That year, Eleanor Roosevelt, who had pressured FDR to put Black pilots in combat, publicized the program by flying as a passenger with a young Tuskegee recruit.

The Tuskegee Airmen would go on to fly a total of 1,578 combat missions and 179 bomber escort missions, losing bombers on only seven of those raids. They shot down 112 enemy aircraft, including three Me 262 jets, destroyed another 150 on the ground and damaged an additional 148. It was a record of mastery, burnished with high commendation.

Yet, for all their remarkable achievements, they were still treated with prejudice. When the fliers trained at the segregated Tuskegee Army Air Field in Tuskegee, Alabama, in the Deep South, they were subject to the harshest bigotry when they stepped off base. They were often called "n—s," and on one occasion Tuskegee Airmen had to give up their seats to Nazi prisoners of war because the Nazis were white.

But Brigadier General Charles McGee said in 2021 that most of the men who flew with the "Red Tails" remembered chiefly their honor in military service, and their pride as pilots. "We were told, 'save American lives.' That's what we went for. . . . It turned out to be a job well done and helped our country change for the better," he said before his death in January 2022. "We were certainly proud of those moments to accomplish successfully what wasn't expected and to help lead the country in equal access and opportunity for all."

The Tuskegee Airmen loved their P-51s like all airmen who took the Mustangs aloft.

One said of his Merlin Mustang, "If the plane had been a girl, I'd have married it right on the spot. Damn right! It was like dancing with a good partner."

The Tuskegee Airmen were so good that bomber pilots constantly asked for them; sometimes there were bitter arguments over who got the "Red Tails"—so named for the distinctive red enamel painted tails of their planes—as bomber escorts. Those arguments could become heated: bomber crews considered the Tuskegee Airmen topflight and everyone wanted them. The Red Tails viewed this as an honor. Their leader, Colonel Davis, had painted on the fuselage of his Mustang the nom de guerre his unit earned well into the war: "By Request."

CHAPTER 20

Oil Run

In the weeks after D-Day, with the great waves of land armies finally breaking out of their beachheads and rolling on to the interior, the fighter war went right on, not slowing or hitting pause for one moment. The mighty Eighth Air Force had teamed with the Ninth to support the land troops moving inland, raking and hammering the Nazi positions. But with the advance uncoiled, as the Ninth gave cover to the troops and divisions below, the Eighth returned to the destruction of German industry. Blakeslee's 4th was ready to go; his group had found D-Day, with the Luftwaffe's failure to appear, a letdown if anything.

The opening salvo was Pölitz. On June 20, bombers accompanied by Mustangs hit the crucial synthetic oil plants there in the easternmost corner of Germany, the biggest bombing raid of the war so far on a facility key to the production of aviation gasoline. To mount the attack, the largest fleet of aircraft ever assembled was harnessed to show the fresh striking power of the Allies. Germany would now bleed oil as if from a gash across its belly.

Until 1944, the oil fields of Ploeşti in Romania had provided almost 60 percent of Germany's oil imports, and they were the chief source of Germany's crude oil. But by August 1944, beginning with the unescorted American raid of August 1, 1943, Ploeşti had been put out of operation. Hitler had planned ahead for a looming domestic oil shortage; by this time, using the so-called Bergius process, Germany was already supplying almost three-fourths of its own liquid fuel needs by turning much of its abundant coal resources into oil. At hydrogenation plants located near the rich natural coal deposits, brown coal was being turned into high-grade gasoline and fuel for tanks and planes. By the summer of 1944, these plants were producing nearly all of Germany's requirements for aviation gasoline. Almost a third of this phenomenal output took place in two huge plants, one at Leuna and an even larger one at Pölitz, in the Silesian coal fields seventy miles northeast of Berlin.

Clearly Germany's oil supply was a prize target for Allied bombers—but in fact, since before D-Day, a debate had raged affecting every bomber flying out of East Anglia. The argument at the top stratum of Allied command would affect every squadron, group and wing of the "Mighty Eighth" and the Royal Air Force; it would determine the focus of bombing across the vast Nazi sprawl to the final phases of the war.

The struggle behind the scenes for control of the bombers had begun with Ike's arrival in England in January 1944 to take over as supreme commander of the Allied Expeditionary Force for the D-Day invasion. As he made his rounds—dining with the prime minister and the king, taking his personal train to inspect troops in southeastern England—the air issue was brewing. It seemed to overshadow everything.

As Overlord approached, tension mounted: What was the best way to cripple the Germans in preparation for it?

On one side, General Spaatz, now commander of the U.S. Strategic Air Forces in Europe, had long held that mortally bleeding Germany of oil would bring her to her end. He was convinced German oil production was the key target. Knocking it out would starve the Germans of their vital fuel, and Spaatz now had the air power to do it. Cutting off petroleum, he said, would drain the Reich's life-blood.

On this, the gruff Spaatz, "untidy and sometimes unshaven . . . with a taste for bourbon and a distaste for being behind a desk," was in direct conflict with the famously difficult Air Chief Marshal Sir Trafford Leigh-Mallory, now in charge of the invasion's tactical arm. Leigh-Mallory—the Cambridge-educated clergyman's son "with a neat toothbrush mustache and the soulful eyes of a well-fed spaniel," arrogant and not easy to get along with—was pointedly at odds with Spaatz. His plan for the destruction of the Third Reich was starkly different.

Leigh-Mallory's master plan called for an attack on the system of French railways leading to Normandy. In advance of the invasion, he drew up a plan for Allied fighters and bombers to strafe relentlessly the rail lines and roads into western France, thereby preventing German troops from reaching the beaches. The architect of the plan was Solly Zuckerman, once described as "a small, mysterious man in an unpressed tweed suit." A South African–born professor of zoology, he had become Churchill's trusted counselor and an advisor to Ike's deputy, the amiable, "pipe-smoking, slim, urbane" Arthur Tedder; Zuckerman would also be the creator— among other inspirations—of the artificial "Mulberry" harbors on the Normandy beaches.

Destroy the arterial ganglia of French railway tracks and the trains that moved on them, and the marshaling and maintenance yards in northern France and Belgium, Zuckerman said, and the Nazis would be gridlocked. No troops or supplies would be able to move.

In March, as the air debate raged on, at times bitter and bloody, Ike's headquarters was transferred from 20 Grosvenor Square in London to Bushy Park, a group of Quonset huts on a private estate near the city. There, in a one-story building with a tin roof, buckling linoleum and flaking paint, Ike lost his patience. He was not feeling well: he had caught a cold in the rainy weather that would not go away. Finally he reached the end of his rope. He issued an ultimatum to his combined chiefs of staff, threatening to resign as supreme allied commander if his chiefs could not agree on a strategy.

"I am tired of dealing with a lot of prima donnas," he discharged. "By God, you tell that bunch that if they can't get together and stop quarreling like children, I will tell the Prime Minister to get someone else to run this damn war." By this dire threat, he got his way. Ike chose "Zuckerman's Folly," as its opponents among the generals called it—targeting rail and transportation for attack.

The campaign against the transport arteries now went into high gear. By the end of April 1944, the Eighth Air Force had flown 33,000 sorties, and railroad targets had been hit by more than 30,000 bombs. Concerned that civilian casualties would affect relations with France after the war, Churchill had set up a system to warn the French population in advance of attacks. During May, Allied pilots claimed destruction of 16,000 freight cars, 900 locomotives, countless bridges and 1,000 aircraft. The French transportation system was severely crippled. Long lines of railcars were now unable to move. Repairs were prevented by more strafing; rebuilt bridges were bombed again. The plan had been called Zuckerman's

Folly, but the result was hardly folly. By D-Day rail traffic toward Normandy was 30 percent of what it had been in January. The Germans could not move their troops.

After D-Day the argument was kindled afresh; every bomber group and fighter squadron waited on the decision that would guide their war. Officially the priority remained transportation. But unofficially Ike gave Spaatz permission to go after oil when his bombers were not needed to support the drive across France. That summer the Eighth began to fly missions against oil facilities north of Munich, the Fifteenth south of Munich and over Austria and Hungary.

In the windy black void of the English night, the 567th Bomb Squad of the 389th Bomb Group was ready before dawn—breakfast at 1:00 a.m., briefing at 2:15 as the bombers were fueled for long distance, then a blessing for the Catholic men from the chaplain. Then, in the great empty pit of darkness, fire up engines; at 4:10, taxi out to the runway; at 5:00, take off—and the B-24 Liberator *Mistah Chick*, along with the other 47 heavies of the 389th, was off the ground, airborne.

Here in the cold, inky winds, the big sky arks rendezvoused with the Mustangs in the wide expanse of altitude, more than 1,000 bombers forming up with hundreds of pursuit planes. Now the planes climbed to 13,000 feet—out over the 30 miles of the Channel, across the flat and vacant water, with the sobering thoughts that fill a man's head in the empty minutes before he goes into battle. "At times I'd say a few prayers to myself as I always do on missions. I'm not scared, but I know that anything might happen in this dangerous game," wrote Lieutenant Veto A. Iavecchia, bombardier in the *Mistah Chick*. "I guess my prayers are for my wife mainly, I worry about her often." The vast armada of sky ships rumbled on, ranging over the broad Channel in the chill altitudes, covering miles of sky. Contrail, cloud. Two or three hours to Germany.

Then they hit the French coast, the Pas-de-Calais and the Nazi stronghold of mainland Europe. They had been told they would meet 200 German fighters and some 200 flak guns over the target—considered intense. Out above the North Sea now, over Denmark. Then, out of nowhere, the tail gunner on the aircraft radio system: "Me 410s are attacking the plane behind us." Iavecchia recounted: "A few minutes later, I saw one of our bombers crash into the sea. It seemed to still be gliding, but then I saw it burst into flame and it crashed into the sea. I could see three parachutes gliding down." Then a shock.

"Suddenly our plane lurched over on its side, and our pilot immediately regained control of the ship; our #3 engine was knocked out. By now the flak was very accurate and intense. I saw two of our own planes nearby diving earthward, probably they got the worst of the flak," Iavecchia recorded. Then, over the target, release bombs—and havoc. German fighters in swarms jumped the Fortresses and Liberators like sharks in a feeding frenzy. "Suddenly I called 'Bombs Away' and I could see my bombs hurtling earthward, twelve 500 pound G.P.s [General Purpose bombs]. I could see huge fires on the ground and billows of black smoke spiraling 1,000 feet into the air. The whole factory site was afire and soon other bombs would play more havoc on the same area," Iavecchia recounted in his diary. The bombing run lasted 12 eternal minutes, the concussions from the shocks rocking the air. Now Nazi fighters veered down through the B-24 Liberators.

"Suddenly 7 Fw 190s hit our formation. They'd open up, throw a few bursts and slide under us and our gunners would be firing away at them." Then Mustangs! "Our pilot called for friendly fighters and suddenly I could see P-51s chasing the enemy all over the sky," Iavecchia recorded. "What a relief to see our boys—and, boy, they're the best in the world." Blakeslee's 4th was in the skirmish, plunging, veering, climbing, "bouncing" lizardlike Focke-Wulfs and stingray Messerschmitts. His team was in the thick of it again, plunging, careening, soaring up in ascent and gunning down to doom.

Lieutenant Donald R. Emerson was leading a section of P-51B Merlin Mustangs when twin-engine bandits were reported; Emerson turned starboard and ran into Me 109s and Fw 190s as well.

"I started after a Me 109 but my No. 4 . . . was able to get into position ahead of me and shot him down. The kite smoking went straight in and no chute was seen." Emerson flicked right past the downed Me 109, and immediately got into a tangle with a Focke-Wulf emblazoned with a red diagonal band on its fuselage behind the cockpit. "There was a silver Mustang behind him [the German] which I cut off in a starboard turn. I was just ready to start pulling deflection when to my amazement he rolled over and the pilot bailed out, his chute opening almost immediately. The impression I got was that the pilot, seeing two Mustangs after him, decided to get out while the getting was good. He may have been hit before all of this took place but he was not smoking and seemed to be under control." One more bogey down.

Mustangs raced, dashed through the air, meeting enemy aircraft. Then-Lieutenant Wallace E. "Lucky" Lowman, another Mustang pilot with the 20th Fighter Group, recalls the Mustang was "faster, more maneuverable. It was easier to fly." When it was supercharged with the Merlin 61 engine, "then it became a real good airplane. It wasn't a tough airplane to fly." Flak puffed and shook; battle flashed past.

"We bounced about 15 plus Me 109s and Fw 190s," said Captain

Otey M. Glass, Jr., of the 4th, in an after-mission report. "The 109 I attacked was making a port turn and I got on the inside . . . firing all the while. Range was about 400 yards and I was closing. He must have seen me as he pulled up in almost a loop. I followed firing and observed good strikes on the right wing very close to the fuselage. As he started down from the top of his loop, he was trailing smoke and was last seen going straight down. I claim one Me 109 destroyed."

The Mustangs hit high, hit low, hit hard.

Lieutenant Charles H. Shilke and his section wreaked vengeance on an airfield near Neubrandenburg:

"Lt. Gillette made the first pass West to East, firing on a Do 217. I followed and fired . . . then pulled up and away when I saw flak coming up from the buildings. . . . I was firing on a Ju 88 and a Do 217," remembers Shilke. "Lt. Gillette returned for a second pass from North to South and fired on a Ju 88. Lt. Gillette saw my [target] start to burn and I saw flames coming from the one he was firing at. Then Lt. Harris came in from West to East and fired at a Do 217." In this one single combat, the section racked up one Ju 88 and three Do 217s destroyed, with one Do 217 damaged.

Mustangs were grabbing victory in the sky everywhere across Germany, on the ground, in the air, shielding the big bombers they covered like cattle dogs, nipping, snapping, flashing teeth at the marauding Germans. The 4th destroyed a total of 15 German fighters at Pölitz that day.

They were rearranging every parameter of the vast clash in the air, upending every convention in the skies over Teutonia and Prussia. In the bitter October of 1943 just before the P-51s entered battle, average bomber losses had been 9.1 percent of the U.S. heavies sent out, with fully 45.6 percent damaged—almost half. As early as February 1944, losses had sunk to 3.5 percent with just 29.9 percent damaged. And the victory of the Mustangs was in boldface: in the one month of March 1944, the Eighth Air Force, led by Mustangs, had destroyed more than twice as many enemy planes as it had in the two years of 1942 and 1943 combined. The Germans were being shot down like clay pigeons in a skeet shoot. That spring and summer, the Mustangs were outflying every type of German airplane.

Major James A. Goodson, the ace of the 4th who flew on D-Day, was leading his squadron in an intense pitched battle over the Neubrandenburg Aerodrome, attacking German planes that were coming in to land after hitting the bombers. He pounced on one Me 109.

"Suddenly I was catching up fast and I realized he was throttling back, entering the circuit to land. It also meant he didn't know I was lining up on him. I waited until I estimated I was less than 200 yards behind him and closing fast. The first burst scored hits all over him. I immediately pressed the trigger again. I was still hitting him when I had to stomp rudder and throw the stick forward and right to avoid ramming him. I looked back and saw him dive straight down to explode on the deck," Goodson wrote in his memoir *Tumult in the Clouds*.

He tagged another 109; then, on the airfield below, he spotted a luscious prize—one of the experimental Me 163 *Komet* rocket planes. Major Goodson attacked at high speed, spinning to avoid flak. He could see his shots slamming into the *Komet*, but ground fire was coming up now.

"At the same moment I felt the plane shudder. I heard the *crump* and smelled the explosive. I felt a numbness in my right knee, and knew I was hit," Goodson recalled. "But it was the plane I felt for. She was like a stricken war horse. She tried to respond but the lifeblood was ebbing. I tried to hold her up, but as I sensed her start to stall, I gently eased the stick forward. And so, tenderly and sadly, I nursed her down, feeling her, caressing her, until as softly as I could, I let her settle on the rough ground. As we hit the ground, I cut the switch. We bumped and skidded to a halt. Suddenly everything was quiet."

He climbed out of his Mustang and stood for a moment on the wing. There was a lot of blood on his right knee and the brown cloth of his pants leg was in tatters. "I looked at my name and the 30 swastikas representing the official victories. I patted the side of the plane," Goodson wrote. Her designation letters were VF-B.

"Two more now, old friend, not bad! We went out on a high note," Goodson recalled saying. At the sound of his squadron above him, he waved at them to shoot up the Mustang, hearing other Mustangs already diving and firing as they finished off his plane.

He stumbled into the forests until he could lurch forward no longer. Slumping down, he looked at his knee and examined himself. His flak boots had protected his legs from the knees down, but the back of his legs above and his thighs were peppered with little bits of shrapnel. His right knee was numb and covered in blood. He probed with his finger and saw a piece of flak had gone right through the fleshy underside of the knee.

He was all alone and badly wounded in the dark green gloom of a forest in Germany, surrounded by thick underbrush and trees. This expanse was not like the gentle beech and oak forests of England, civilized and domesticated, but rougher, a wilder expanse of dim shadow. Four deer went bounding past. He knew nothing about the land or where the thick forest was.

After an extraordinary journey through several jails and detention by the Gestapo and the SS, Major Goodson would wind up in a German prison camp. He would survive the worst and remain alive, but would sit out the rest of the war as a POW.

High above Pölitz, the concussive fire and flak went on as the Eighth
Air Force battlewagons kept dumping their bombs on the petroleum facilities, and the Mustangs continued their unrelenting duel
with the enemy vultures. Colonel Chester B. Hackett, a B-24
bomber pilot in the 389th Bomb Group, now homed in on the target, too, an approach that would turn into a spectacle. Hackett
asked the group commander, Colonel Jerry Mason, whether they
should change their altitude to avoid some of the flak; Mason said
no, they must stick to a steady track. Flak bursts, black, brown. The
389th sky rams thundered on, their Pratt & Whitney radial engines
howling. "We made our run at the same altitude as the formations
ahead of us." Next a sudden, massive shock.

"Just after we dropped our bombs, my aircraft seemed to stop in
midair. We had taken flak hits in the nose section, bomb bay, fuel
tanks and waist compartment." The mission was over for Hackett
and his crew.

"Dropping out of formation, I called Col. Mason and told him I
was going to try and reach Sweden," Colonel Hackett wrote later.
"Power on all four engines was just about nil, so I called the crew
and directed them to bail out." The 10 airmen began jumping from
their stricken ship. All at once, one of the team called out that a
lieutenant had been hit and was badly hurt. Hackett told the crewman to attach a static line to the man's parachute, shove him out
the waist window, pull his rip cord and jump out himself.

Smoke began filling the cockpit. Hackett prepared to jump,
alone now with the plane's bombardier, Captain Garland East,
whose parachute had been destroyed in an explosion. Captain East
advised Colonel Hackett to jump—he would try to set the plane

down himself. Hackett refused and began his descent for a crash landing.

At about 15,000 feet, from a clear blue sky, two Me 109 Messerschmitts appeared and came in to attack.

Colonel Hackett now put the bomber into a steep dive, picking up speed, turning in to the enemy fighters, splitting them apart. One crashed to earth. The determined pilot saw a grain field; at about 300 feet, he lowered the hydraulic landing gear and set the fragile Liberator down on German soil. "There was very little wind and we hit at a fast rate of speed. After rolling about 2,000 feet we hit a ditch and snapped the nose wheel," he wrote later. The plane came to a stop.

Out of nowhere in the sky overhead, two Mustangs appeared. The P-51s had shot down the Me 109 Hackett had seen crashing.

Hackett and East were down in Germany, somewhere in the north, alone behind enemy lines, stranded and in danger. Later they were captured as prisoners of war. The two Mustangs had undoubtedly saved their lives.

Other P-51s ravaged other sectors all across the wide northern expanses around Pölitz near the Baltic Sea coast that Tuesday, June 20. A team of Mustang marauders ran into more than 25 Nazi fighters near Griefswald Bay at 27,000 feet; then another gaggle of five or six Me 109s at 15,000 feet swung around to attack them. "Lts. Monroe and Dickmeyer destroyed two of these Me 109s. Several e/a were seen on the deck. Captain Joyce, Lts. Cwiklinski, Malmsten and Gillette attacked these e/a. . . . Lt. Gillette then joined [and] . . . destroyed a Ju 88 and damaged two Ju 88s on the ground," one report stated. The 4th Fighter Group in this one action racked up three Me 109s destroyed in the air, two Ju 88s destroyed in the air, one Ju 88 destroyed on the ground and one Fw 190 damaged in the air. Full strike in the clouds.

Another section of P-51s over the Lübeck area ran into a wolf pack of 50 Me 210s, 110s and 109s. The two swarms tangled "and our pilots got five claims out of the deal with a total of three destroyed. . . . This was a good half day's work," one after-battle account remarked.

Back inside the massive aluminum fuselage of *Mistah Chick*, Lieutenant Veto Iavecchia was almost done, winging away, flying through the blizzard of fighters, antiaircraft fire and pillars of smoke rising into the air, ignited by the 12 "bricks" he had dumped along with the other bombers. Back across the interior, course almost due east. Finally they came in and made their way home, finishing a splendid run that had ended—for some with triumph and others with bare survival. For Veto Iavecchia's crew, it was touch and go.

"About two minutes past the target, I felt the plane sink about 1,000 feet in about 15 seconds," Lieutenant Iavecchia recalled. The B-24 was hit, lost its number three engine and was streaming gasoline from its tanks.

The gas gauges registered empty. *Mistah Chick* kept on flying on three engines. "I knew now that we'd never get back from this raid," Iavecchia remembers. The pilot, Colonel Ralph Leslie, drew a course with his navigator for neutral Sweden.

Over Denmark they jettisoned anything they could to lighten their load—.50-caliber ammunition boxes, radio equipment, the bombsight, the navigator's bag, anything they could grab.

"I could see the coast of Sweden and what a swell feeling, for I dread the thought of bailing out over water," Iavecchia recalled.

Colonel Leslie presently spotted a private airfield—very small and not meant for a giant aircraft like a heavy B-24 Liberator. They began to glide in, losing altitude. The waist gunner reported that the whole waist of the ship was covered with about four inches of gasoline. By now they were only 100 feet off the ground. The *Mistah Chick* cleared a sandy beach by about 20 feet, glided in, dropped and, with a deafening groan, hit the earth, plowed across a field, lurched through the loam and finally came to an unceremonious stop.

Iavecchia and the others scrambled at once to safety; the lieutenant stood dazed on the ground, just staring. He was just staring. Two men came running, racing up to observe the spectacle of the great ruined leviathan. Iavecchia blinked.

"I hurriedly asked, 'Where are we?'"

"Sweden," one of the men answered.

And then Veto Iavecchia did something unexpected: he threw his arms around the Swede and kissed him. The man smiled.

Soon soldiers came up with rifles and bayonets drawn; while Sweden was technically a neutral country, it was occupied by Nazi Germany. The American airmen were taken and searched for weapons.

Later the men were hosted by the Swedes for a grand dinner and a night at the best hotel in town. Finally, they were taken to a temporary internment camp as prisoners in a neutral country.

Some 1,400 bombers took part in the Pölitz raid. Seventy-two failed to return; two planes crash-landed in Sweden. With Mustangs on the raid all the way over the target, the loss rate was now down to 5 percent. Thirty-four German planes were shot down. Despite taking a mauling, the bombers of the Eighth were able to cause substantial damage to the synthetic oil plant at Pölitz, all made possible by Mustangs—with their remarkable ability to outperform Germany's fighters and the phenomenal range that allowed them to fly distant missions.

The Mustangs were raking in winnings everywhere across the sky. The June 20 raid on Pölitz was a good start; a second strike would follow. This would be important. The plant—manned by forced laborers from the camps at Stutthof, Sachsenhausen and Ravensbrück, 13,000 of whom would die there of hunger and disease—made 15 percent of Germany's synthetic fuel: gasoline, oil and lubricants. Because of the raid on Pölitz and others like it, by early July German oil production had decreased by 75 percent.

Soon the military would be down to only five days' supply of oil for heavy operations. German planes would be towed to airfield runways by horses to save fuel on taxiing. Without the fuel the Pölitz plant and others produced, the Nazis would soon be unable to win the war.

Selected Targets In
The European Air War
1943–1945

SELECTED CITIES AND AREAS
TARGETED BY AIR RAIDS

CHAPTER 21

A Perfect Show

The very next day, the weary pilots of the 4th rallied for a highly secret and sensational run that had been long in the planning. The air force had come up with an almost outlandish scheme—code-named "Operation Frantic"—that would send American wing power to points in Russia and Ukraine previously beyond reach. This would be a record distance for a single-engine fighter, and an astounding marker of how far the P-51 could fly. The Mustang's vast range was the crucial factor in the gambit.

The scheme would launch bombers, escorted by Mustangs, from their bases in England to isolated German objectives, then on to Russia to land at three Ukrainian bases—Pyryatyn, Poltava and Mirgorod. From there the fighters and bombers would hike a second, southbound leg to Foggia, Italy, bombing objectives in the Balkans along their way. Finally the intruders would make their way home to England for landing, after a third round of bombings en route.

It was a three-sided triangle: England to Russia, Russia to Italy, then Italy back to England—and it would afford the Eighth and Fifteenth (based in the Mediterranean) three windows for bombing in one single journey.

Operation Frantic was a bold gambit: the first leg from England to Russia was almost 1,500 miles long and would involve 7.5 hours in the tight single seat of a P-51. But Mustangs had flown eight hours before: the prospect was within reach. This elaborate relay became known as the Russia Shuttle. It would show the Nazis once and for all the prodigious capabilities of the Mustang. It would be a radical experiment and, if it succeeded, an aviation milestone.

One of the first Mustang outfits selected for the daring mission was the 4th Fighter Group. It was the kind of audacious gambit Don Blakeslee loved, yet even he was unsure it would work. Group Intelligence had been up all night preparing final course maps for the mission. But none of the maps had Russian landmarks on them, and in any case the weather would be overcast all the way to Pyryatyn. The ground would not be visible through the cloud cover. Radio silence would be strictly observed. Blakeslee had collected no fewer than 16 maps to take in his cramped cockpit, but in the end he would have to rely on his compass and dead reckoning to guide his men from Debden across a distance half the breadth of the Atlantic Ocean to a speck in the middle of the Russian tundra.

Rumors about the mission had flown around the base since several days before, when the medics had lined up the pilots and crew chiefs to inoculate them against cholera and other exotic bugs. Now, on June 21, the day of the mission, Blakeslee briefed his boys. The War Department had ordered a secret film made of the briefing. Wearing a white scarf for the occasion, Blakeslee strode into the briefing room, pointed to Pyryatyn on a huge map, and said to the assembled men, "I want you to land 68 aircraft at this place."

"Now look," he said. "Before we all get excited about it, I'll say the whole trip is about seven and a half hours. We've done 'em that long before. We'll be throttled back, so Christ, we could stay up for eight hours. . . . On the way to Russia . . . you will not drop your tanks. . . . If for any reason you should have to drop tanks around Berlin—you've had it. You'll have to return to Debden. . . . Once we make rendezvous with the bombers, there will be absolutely no radio conversation. If you see a man's wing on fire—just be quiet, he'll find out about it himself after a while. . . . For Christ's sake, no landing errors. The Russians shoot the men who make mistakes. . . .

"Let's make a pretty landing . . . a pansy landing . . . bang, bang, bang. We want to make the thing look like a seven and a half hour trip is nothing to us." The point of the mission was to demonstrate that Allied bombers with escorts could shuttle at will from England as far as Russia. It was calculated to rattle the Germans. "This whole thing," Blakeslee said, grinning as he finished the briefing, "is for show."

All the ground personnel gathered in the rain to watch the 68 Mustangs take off. Colonel Blakeslee and his second-in-command were the first to clear the field. They took almost the whole runway to get aloft: their heavy drop tanks weighed 108 pounds. They raced down the runway, lifted into the air, devoured the altitude as they climbed over the fields and cottages of Essex and mounted into the cloud cover. Then, flying at the head of the group and still ascending into the dome above, Blakeslee set a course in the overcast sky for Russia and led his pilots east toward the sea.

Airborne, the Mustangs joined their wing of Flying Fortresses rendezvousing aloft—a bomber force of 104 battlewagons. Complex safety precautions had been taken for the risky ploy: a vast diversionary armada of 1,000 bombers had been sent to Berlin as a decoy. None of the fighter planes were to engage in combat. The pilots had all been instructed not to carry guns—if they had to bail out and Russian partisans came upon them armed, they would be shot. Now they flew on toward Berlin, over the North Sea, over the slate gray traveling swells, the coast of Europe, watching, homing in on Ruhland, their target, 50 miles from the Polish border.

South of Berlin, heavy flak came up. It followed them all the way in, brown, black, gray puffs of ground fire filling the last miles to target, exploding black clusters. They homed in, staying steady, level, straight as they must for accuracy. Bombers couldn't evade, dodge or dive. One Mustang was hit. All at once they were over the drop zone; the bombers loosed their steel jugs over the MPI (Main Point of Impact), a synthetic oil plant at Ruhland and nearby rail yards at Elsterwerda. Then fighters. From nowhere, 15 to 20 black-nosed Messerschmitt Me 109s pounced on the bombers, making a hit on the egg baskets. The Mustangs chewed into the Germans, tore apart five and shot them down; one Mustang was lost. Finally they were clear and away.

They flew on into the dying light of the late afternoon, Blakeslee and his "Blakesleewaffe" howling east toward Ukraine, over Poland, Warsaw and Kyiv, toward a speck in the vast Eastern European wilderness. Minutes seemed like hours. They could see nothing below them; it had been overcast and now it was darkening, too. Blakeslee nervously checked his watch: 7:15. Fuel was running low. He looked at the schedule strapped to his leg, and at his maps. They should be arriving in 20 minutes. They descended, approaching the navigational plot where the base should have been, flying low to the ground, coming into landing configuration. Still nothing. Cloud. Sunset. Endless sky. Hundreds of miles of the Poltava Oblast region of Ukraine stretched away. Silence over the radio. The Packard Merlin V1650-3 engines roared through the dusk, never slowing through the hundreds of miles receding behind them.

Then a pattern of signal flares in the dusk.

The lights outlined the field below at Pyryatyn; Blakeslee had hit the bull's-eye to the minute, 2,000 miles from their takeoff at Debden, deep in the heart of the remote Eurasian landmass. They were scheduled to land at 7:35; it was 7:35 exactly as Blakeslee's wheels hit the tarmac.

His wingman, Lieutenant George H. Logan, Jr., saw him fling his maps in the air. Blakeslee turned and blew kisses to Logan from his cockpit. Over the radio, the gruff ace of countless combat hours yelled: "The end of a perfect show!"

One by one the other pilots dropped onto the landing field and rolled to a stop. Russians climbed onto the wings of Blakeslee's Mustang and presented him with a bouquet of red roses. Then he was hastily snatched away from the field to be rushed to Moscow for a radio broadcast. The mission was going to be big news in America. With his customary aversion to the press, he would later pronounce the interview "tougher than the trip over."

The other pilots climbed out of their cockpits and stepped onto the wings of their Mustangs, surrounded by cheering Soviet soldiers. Russian officers drew up in a car. Howard "Deacon" Hively, an ace with 12 victories from Athens, Ohio, who was one of Blakeslee's close buddies, spoke in broken phrase book Russian, asking for something to drink. He wanted water or soda, but the Russians barked back, *"Da, da. Schnapps. Schnapps?"* Vodka.

"Da, da," said Major Hively. They went directly to the local hospital to wash up, Hively with the Russian commanders in an old Ford. Then they were taken to a banquet hall where they drank shots of vodka and toasted Marshal Stalin, President Roosevelt and the damnation of the Nazis—with fiery jiggers each time. His hosts brought Hively another bouquet of roses, kissed him on the lips in the Russian manner and danced, men and women with the

American pilots according to Russian tradition. A balalaika orchestra rang out dance music.

Later Hively was received by a Russian general. More vodka was downed. Hively pulled out an American five-dollar bill and signed it; the general pulled out a Russian 100-ruble note, signed it, too, and handed it to Hively. More vodka, more toasts, more vodka. That night Hively was given the commander's bed to sleep in. The Russian airmen and the American fliers were fast friends.

Jubilation had surged over the Americans and Russians at Pyryatyn, but the next day disaster intervened amid the giddiness of victory. That night the B-17s, which had landed at Poltava, and the Mustangs at Pyryatyn were attacked by German bombers and fighter-bombers. Undetected, a lone Heinkel bomber had followed the American formation out of Germany and tailed it to its landing in Russia. The next night the Germans swooped in, pounding the bases at Poltava and Pyryatyn. Forty-three bombers were blown up at Poltava; 15 Mustangs were put out of action at Pyryatyn. A Russian Yak fighter plane shot down one Ju 88. In the dark the Mustang pilots jumped into action; they got no hits but sent the Germans scrambling away, not to return.

After five days of maintenance by the Debden mechanics, who had flown in with the bombers, and hearty, ample brotherhood with the Russians, the Americans pulled out to fly the next leg of their journey, again by compass and chart, to Foggia, Italy. On their way, the swarm of Flying Fortresses and Mustangs hit an oil refinery in Poland, pounding the installation with a saturation of bombs. The Germans never attacked and the "Fourth But First" 4th Fighter Group passed off its bomber convoy to Mustangs of the Fifteenth Air Force. The Blakesleewaffe landed safely at the Lucera Air Force Base at Foggia, completing its second leg.

Deacon Hively got to see his brother, an Army Air Forces sergeant posted in Sardinia. The pilots of the 4th got to swim in the warm, calm waves of the Adriatic Sea for several days, enjoy its soft, sandy beaches, visit Rome and its ancient ruins and enjoy Italian cuisine—Chianti and antipasto—and the hot Italian sun. Then, on July 2, they got to prove their worth.

That summer Sunday they were to join wing tabs, flaps and ailerons with their fellow pilots in the Fifteenth Air Force in a joint strike near Budapest. Blakeslee's team sortied, climbed into the sky and promptly lost several planes forced to abort because they were ill-equipped for the warm, humid Mediterranean skies. The 4th roamed on nevertheless in reduced numbers into Hungary, venturing on into the bristling fortress of occupied Europe. The Packard Merlin V1650-3s thundered ahead, escorting their flotilla of heavies to Budapest, closing in on ground zero—when suddenly the Luftwaffe jumped. The 4th called on the Fifteenth to assist, but the calls went unheard or ignored. Blakeslee's boys were on their own. Shot. Salvo. The scream of engines. The clatter of machine-gun fire. Hot sun. Lieutenant Donald R. Emerson now found a Messerschmitt Me 109 that had penetrated the 50 or so Mustangs in formation— and attacked. The Messerschmitt climbed at once into a steep ascent, with pieces of the plane flying off. Emerson's Mustang stalled and spun away; the 109 came flying past with one wing shot off. One bandit down.

Now Blakeslee picked off a Boche. Captain Frank Jones of Montclair, New Jersey, bagged one himself, but his buddy Lieutenant George I. Stanford was shot down, plummeting earthward. Across the glare of the sky, Captain Sheldon W. Monroe's group charged into a field of some 25 aircraft—and dispersed them like a flock of scattered vultures.

Deacon Hively locked on another 109 and squeezed the trigger of his guns, top of the stick, firing all four machine guns at once, smoking away at his quarry. Another Messerschmitt was simultaneously attacking Hively while at the same time being ambushed himself by Lieutenant Grover C. Siems. Then another Mustang was tearing into the Messerschmitt on Siems, and a third Messerschmitt was soon savaging this third Mustang. Fighters whirled by in seconds. The sky was filled with thunder, smoke and fire. The Me 109 Deacon Hively was bouncing blew up after a quick burst of .50-caliber rounds—but the Messerschmitt behind Hively hammered him. One round detonated his cockpit canopy; his right eye was injured, the vision in his other eye dimmed by blood. He briefly lost control of his plane but recovered and shot down two more Jerries.

Elsewhere in the veering, plunging battle, Lieutenant Siems was in action. A cannon shell slammed into his shoulder and neck. In excruciating pain and disabled by his wounds, Siems managed to get his Mustang back to Italy and land safely at a bomber base.

In the action of July 2, 1944, all bombers deployed over Budapest by the Fifteenth Air Force were protected by the shield of the 4th Fighter Group. Blakeslee's 20 Mustang warriors alone, without the help of Fifteenth fighter support, successfully held off some 60 Messerschmitts. Blakeslee's stable had added to its burnished reputation as "Fourth But First."

On July 5, they were homeward bound on the third leg of their circuit, the group having now passed the milestone of more than 600 enemy aircraft destroyed—an outstanding record. Over Corsica, the Blakesleewaffe rendezvoused with bombers and hit railroad marshaling yards in France along the way. Then they passed their Big Brothers off to the 56th Fighter Group, the "Wolfpack," flying the shorter ranged P-47s over Châteauroux, France. Spent, weary, bearing the toll of their dead and wounded, they were finally coming in over the Channel on the long glide slope into their airfield, the rays of gentle sun slanting low in the sky, vectoring into Debden over the sprawling green countryside of Essex, draining to Southend-on-Sea and the harbor at Sheerness.

Blakeslee had radioed ahead, and the whole base raced out to watch the adventurers land. The Mustangs came in low over the hangars, but those on the ground noted there were none of the usual aerial stunts—just touchdown on the quiet English fields.

Operation Frantic, the Russia Shuttle, continued for three months. Seven Mustang groups flew missions to Russia, five from the Eighth Air Force and two from the Fifteenth, including one mission by the 20th Fighter Group from its base at King's Cliffe. The last mission was flown on September 18, 1944. The repetitive hammerblows of Frantic brought raids on, among other targets, Buzău, Romania; Mielec, Poland; Ploeşti and Focşani, Romania; Gdynia, Poland; Toulouse, France; and Chemnitz, Germany.

American airpower had reached inside the vast armored sprawl of the Nazi estate and hit where no Allied bombs had ever fallen before. But the point was not the bombs dropped or the German planes shot out of the air. When Blakeslee told his pilots as he briefed them in Debden, "This whole thing is for show," he was not wrong. The Russia Shuttle missions showed Hitler that no corner of his empire was safe from attack by air. The inspiration of Edgar Schmued and the vision of Tommy Hitchcock in the hands of Don Blakeslee had convincingly seized the skies. As Churchill put it in 1944: "Hitler did make Europe into a fortress, but he forgot the roof."

CHAPTER 22

Closing the Ring

The Germans never let die their ferocity or fanaticism. The heat of their fervor, coupled with the bulging aircraft production of the Nazi arms minister Albert Speer, kept the Nazis' air war alive. On the ground, Hitler continued to hope he could form a new front in France and drive back the Allies; he reinvigorated his armies with a fresh draft of all men between the ages of 16 and 60, with troops returning from the Western Front and by strategically shifting reserves.

In the long, hot summer of 1944, flying from bases tucked into the fertile fields of eastern England, out over the traveling blue waters of the North Sea, the Eighth Air Force took off again, resuming its schedule. The fight in the air was close fought and costly. While the Germans lost more bitterly, the Eighth took casualties, too, on the long road to Berlin. In the month of June, the Eighth flew on 28 days; 27 days in July; 23 in August—a draining pace. During that period, it lost 1,022 bombers, almost half its strength, and 665 fighters to flak and the still dangerous fighters of the Luftwaffe. All these casualties were replaced by new arrivals from America. But the Germans were being defeated in the air at an even more disastrous rate by the flocks of P-51B Merlin Mustangs. In the summer of 1944, the *Jagdgeschwaders* of the Luftwaffe were losing 300 planes a week.

The Luftwaffe sent its menacing squadrons aloft in an unceasing struggle to halt and destroy American and British bombers, forming *Sturmgruppen*—storm groups with armadas of up to 40 planes, among them the monstrous new aircraft dubbed the *Sturmbock*, or "Battering Ram," an Fw 190 that had been completely remade and heavily armed. But attrition was the blight that drained the Luftwaffe now as pilots perished in battle at an increasing rate. The German system was to require pilots to fly until they were killed or disabled; the USAAF rotated its airmen out of combat after 35 missions. The cruelty of the Nazi system produced some remarkable, anomalous results: while the top American ace, Major Richard I. Bong of Wisconsin, racked up 40 kills in the Pacific, and the leading Mustang ace, George E. Preddy, Jr., chalked up 26.5 victories in Europe, there were 107 German aces with more than 100 kills, 15 with more than 200 and two with more than 300 kills. It made for a barbaric relentlessness in the career of a German flier. The arrangement also produced the gruesome spectacle of student pilots drafted into fighting—the Germans called the phenomenon *Kindermord*, or infanticide. Luftwaffe duty was one more of Hitler and Göring's fantastic cruelties.

The atrocities at the hands of the Nazis continued unchecked in this last haunted year of the war. Hitler told German civilians they should be willing to die for the Reich; adolescents were drafted and sent out to war on land just as they were in the air. The Nazi internal

235

security services—the Gestapo, the SS and the SD—continued their dark, draconian clasp on all German life.

And the Germans continued their religiocide against the Jews, eventually exterminating half the world's Jewish population, some six million people, in their concentration camps. On January 27, 1945, the Soviet Army would liberate Auschwitz, one of the largest Nazi death camps, where between 1.1 and 1.5 million people, 90 percent of them Jews, had been gassed naked in family units and groups. The Red Army would find only 7,650 starving, sick and freezing prisoners alive.

There had been charged debate in 1944 over whether the Allies should attempt to bomb the railroad tracks leading into the camp or the ovens of the crematoria at Auschwitz. Two Jewish prisoners who escaped the camp that June had described the atrocities inside; their harrowing testimony reached U.S. and British officials through the Jewish underground. Churchill himself urged an attack on the installation, but both the British secretary of state for air and U.S. officials declined, citing, in addition to risk to the prisoners, distance from Mustang bases in England and the impossibility of an escorted mission. In fact, the P-51s had recently flown close to Auschwitz on their way to Pyryatyn in June and Mustangs from the Fifteenth escorted bombing runs over southern Poland that July and August. The Mustangs had the capacity to take bombers into the region. But the mission never went ahead.

Through the summer and fall, as Spaatz's oil campaign took off, the Allies rushed on from Pölitz to hit oil targets like a hammer slam, over 500 air strikes in all. The U.S. Eighth and Fifteenth Air Forces hit synthetic oil fields, while the RAF concentrated on the rich industrial valley of the Ruhr.

The Americans would make 347 strikes on oil plants and the RAF another 158. The June 20 raid on Pölitz had been a massive, crippling blow; it kicked off a campaign that would spill oil from Nazi Germany's system like plasma. It was an assault Hitler's arms minister Albert Speer had mortally feared. He increased his anti-aircraft force to nearly a million strong, setting up *Grossbatterie* of 36 guns each at these critical facilities. In the summer and fall oil raids, the Eighth Air Force would lose twice as many bombers to flak as to fighters.

The assault hit a vital artery, as critical to the Nazis as man-power. Oil was crucial not only for fuel but also as a lubricant for all the flying, sailing and ground machines of Germany's modern war-fare. In October and November, Eighth Air Force bombers inserted by Mustangs continued to sweep across Germany, ravaging oil refineries, tank farms and oil fields in the north. The campaign reached its climax in November, when 37,096 bombs were dropped by the American and British air forces. German oil supply dropped to 300,000 tons, just 23 percent of what it had been six months before.

Spaatz's theory had proved right. Germany never lost its war for lack of hardware or equipment; by the end of the war, Speer's pro-duction phenomenon was turning out 30,000 tanks per year. In the fall of 1944, he had built the Luftwaffe back to 3,000 planes. By 1945, air strength was 5,824, greater than the 5,409 planes that Germany had started the war with five years before. But these weapons could not be used without gasoline. By the end of the war, oil production was 5 percent of what it had been the previous year. The Wehrmacht had enough gasoline for only a few days of heavy operations. Oil for the lamps of the Nazi hydra had run out.

In the last phase of the war, transportation was hit again. At an October meeting of the air chiefs, Spaatz's oil offensive was officially given priority. But the generals agreed that transportation would be hit when weather prevented visualizing oil targets clearly enough to bomb accurately. As it happened, the weather over Germany was overcast from October 1944 until early 1945. In the last three months of 1944, 80 percent of the Eighth's missions were flown in moderate to dense cloud cover. By default, with visibility poor, the Eighth dropped more than half its bombs on transportation targets. Through the fall and winter, American bombers went after the rail marshaling yards, located in towns and cities and surrounded by residential areas. In the heavy fog and mist, the bomber squadrons, which could see nothing on the ground, would "bomb on the leader," the lead plane, which alone carried radar. The leader would hit the target, the others the rest of the town. As one bombardier put it: "There were about 400 bombers ahead of us as we approached the target. Our shadow, if we had had one, would have covered almost the entire town. . . . The planes at the head of the formation wasted the marshaling yard, but the rest of us wasted the town."

Through the fall, Allied bombers, with occasional help from the Mustangs, struck at a wide array of rail and water channels. By D-Day, railroad activity across occupied France had been reduced to 30 percent of what it had been in January 1944; by July it was 10 percent of January levels; by March 1945 freight from the Ruhr Valley could no longer be moved. In the end the Wehrmacht would be paralyzed. Wrote one historian, if the Germans still produced "they could not haul."

By the fall, the Flying Fortress–Liberator-Mustang team in tandem had brought low the German giant with its one-two punch. Big Brother and Little Friend had all but subdued the Prussian monolith. Interviewed after the war, German generals concluded that three things had defeated them in the end: the loss of an oil supply to fuel their phalanxes; the destruction of transportation to move reinforcements to the front; and the Mustangs, which had taken bombers into the center of Germany. The Mustangs had cast their net around the entirety of the Nazi war machine.

"The P-51, especially in regard to the strategic bombing campaign over Europe, was a war-winning weapon," says Jeremy R. Kinney, associate director for research and curatorial affairs at the National Air and Space Museum of the Smithsonian Institution in Washington. "It helped turn the tide and make the idea of strategic bombing work in regard to having a fighter airplane capable of going all the way to the target and back to make sure that these heavy bombers could hit their targets. And they did."

From Paris in August, the Allied siege rolled on through the vineyards, orchards and ancient villages of the Western Front. The American-British-Canadian drive across 900 miles of France, Belgium and the Low Countries came to a halt in September at the Siegfried Line, the long, curling 390-mile north–south chain of pillboxes and strongpoints running from Alsace to Stuttgart, with more than 18,000 bunkers, tunnels and tank traps guarding the western border of Germany. Six Allied armies drew up that fall and halted in the rolling hills and thick forests of the western frontier to regroup and reorganize. Their long victorious thrust had substantially thinned the American, British and Canadian ranks and rows; lines of supply from England and the rear areas in France were strained. Generals Patton, Bradley and Montgomery were short of ammunition, medicine, food and guns. Except for Cherbourg, the Germans still held the major French ports along the Channel. Most of the supplies for Allied troops had to come across the Normandy beaches. And with the French railways decimated by the Eighth Air

Force before D-Day, supplies had to move by road. The advancing troops were dependent on the "Red Ball Express," a truck service by which, 24 hours a day, west to east, more than 6,000 trucks and 23,000 men brought food, gas and ammunition for the Allied armies from the beaches and the port of Cherbourg. Until late November, when Montgomery's troops reopened the captured port of Antwerp, the Red Ball Express would be critical. ·

The commanders held up while their troops gathered strength.

The lull suited the Germans, and in the slow weeks of summer's fade into early fall, they used the pause in the Allied drive east to bring up their reserves. The Germans amassed strength more quickly than the Americans: by late fall, Nazi commanders had 24 divisions in the west under the veteran field marshal Gerd von Rundstedt, who had played such a large role in the defeat of France in 1940.

Thus, in autumn, as the rain and cold weather dampened morale, both sides held back, reinforcing, summoning strength, pooling and collecting the power of their forces in the field, preparing on both sides for a new onslaught of Eisenhower's advancing drive. The Führer was putting the finishing touches on a plan for a last heroic effort to reverse the course of the war. In late July, members of his general staff had planted a bomb in Hitler's headquarters, the Wolf's Lair in East Prussia, in an attempt to assassinate him. Suffering from shock and plagued by pierced eardrums in the aftermath of the explosion and by fear that he would again be betrayed, he holed up in his bunker. He had his orderlies spread out maps of the Ardennes at the foot of his bed: in December he would launch a great counterattack. He would replay the triumph of 1940, when his Panzers had invaded France through the same thick forests. In the dense woodlands of the Ardennes, the Germans had brought up the Fifth Panzer Army under General Hasso von Manteuffel, reinforced by the Seventh Army, and the Sixth Panzer Army commanded by SS General Sepp Dietrich. By starving the rest of the front of gasoline and ammunition, Hitler made sure they would have sufficient cannon fodder for a long battle.

Now the cold set in as November gave way and the armies coalesced, both the Germans and the Allies hesitating, pausing, holding in suspended animation as both sides drew strength, and all the drawn battalions and brigades playing a waiting game in the sharp of the chill.

CHAPTER 23

Buzz Bombs and Doodlebugs

On the night of June 12, 1944, Churchill, after visiting the Normandy beaches to view for himself the progress of the landings, arrived back in London just before midnight. As his train neared the city, the first specimens of a terrifying genus of new German rocket were being launched from Belgium and northern France. Flying at 350 miles per hour at between 3,000 and 4,000 feet, they were jet propelled and carried a 1,875-pound high-explosive warhead. These were the "buzz bombs" or "doodlebugs," as Londoners called them, creations of Hitler's advanced wonder-weapons program—the *Vergeltungswaffen*, his desperate attempt to turn the tide of the war. Sirens filled the air day and night; Londoners slept in the subway tunnels once again. The V-1s that hit Britain in the next four weeks killed 2,752 and destroyed more than 8,000 homes. The siege continued for months. By September, 2,400 of the buzz bombs had reached Greater London, killing more than 6,000 and destroying some 25,000 homes.

Beyond the deaths and the damage, though, the buzz bombs caused constant and unbearable strain. Most Londoners felt that the V-1 siege was worse than the Blitz. As the skies emptied of V-1s by September 1, 1944, Londoners heaved a sigh of relief. But relief was premature; things would only get worse. On September 8, a much deadlier threat arrived. Two V-2 bombs streaked across the English Channel from the Hague in the Netherlands and fell on the city. This had been rumored for months, as intelligence had suggested that Hitler was planning an even more intimidating weapon from the air. Where the V-1 was the size of a small fighter plane, the more advanced V-2s were 46 feet high, weighed 14 tons and had a top speed of 3,600 miles per hour. The warhead on the V-2 could wipe out a city block. Most unnerving of all, the V-2s could not be heard falling. Only when they were close to the ground and detonated did they give out a roar like a freight train. By October the attacks were constant, and they would continue until March 1945. All told, 1,115 V-2s would be fired, the majority at London; 2,754 Londoners would die, with an additional 6,523 injured.

In May 1943, high-flying planes had photographed more than 100 baffling construction sites in northwestern France. Obviously, the gigantic sites were for missiles. Photo analysts could see the truckloads of concrete being poured into the installations. Measurements showed they were lined up on London. Allied intelligence realized there were eleven launching sites for the missiles, including six for the V-2. That November the first of 96 "ski sites," or launching sites—so called because of their raised launching ramps—for the V-1 rocket were also seen in France.

The RAF carried out the first raids in August 1943, and in November a joint British-American operation was launched against the rocket bases under the code name "Crossbow." In mid-April Churchill and the British War Cabinet, nearing desperation over the destruction caused by the V-1s in London and the evacuation of a million women and children, pleaded with Eisenhower to devote more Allied planes to destroying the buzz bomb sites. Spaatz and RAF air chief marshal Harris protested vehemently: Operation Crossbow would be a huge distraction from the urgent preparations for D-Day. But with intelligence reports showing the V-1s would be followed by V-2s, from mid-April 1944, 40 percent of bomber sorties against German industry would be diverted to attack the V-1 and V-2 launch sites.

By the late spring, 24 of the launchpads had been destroyed and 58 seriously damaged. Importantly, the steel-and-concrete rocket bases—the so-called No Ball sites—sustained enough damage that the first nightmarish V-1 attacks were interrupted and delayed for three to four months. By the time the first V-1 fell on London on the night of June 12, 1944, the Normandy beachhead was well established.

By bugging the telephone calls of a high-ranking German prisoner of war, British intelligence then learned about the top secret long-range rocket research at the Peenemünde Army Research Center in Germany. With the aid of sketches and maps of the facilities made by two Polish janitors at Peenemünde and information from the resistance group around the Austrian priest Heinrich Maier, later executed by the Nazis, the Allies were able to launch air assaults on the closely guarded Peenemünde installations near the Baltic coast.

The August 4, 1944, American raid on Peenemünde would be the third of four Allied attacks on the research base; it would slam the experimental station hard. On that day, Mustangs of the 20th Fighter Group sortied again into the skies to fly escort. The Mustangs knocked off Nazi planes like carnival pins. They destroyed 23

German fighters of various types and bagged nine on the ground. They hit Peenemünde at 2:27 in the afternoon.

The bombers came in over the wide blue-gray flint seas of the Channel, configured into their interlocking combat boxes in the hot honeyed sun. They cleared the English coast at 11:07, the enemy coast at 1:18. Then in over Jutland—Hamburg, Rostock, Schwerin, Lübeck—at 22,000 feet, then up to 24,000 feet, straight, no waver in the flight path, out to the Baltic shore, due east. Now they came above the target, drew ahead over ground zero and dropped their eggs. "Bombing results were observed to be good to excellent just a little over the MPI," recalled Lieutenant Colonel Robert S. Kittel of the 379th Bomb Group. "Our lead group got numerous direct hits on the target and one extremely large explosion, accompanied by flame, was seen," recorded another 379th Bomb Group officer. For minute after strained minute, the heavy sky ships drew ahead like a column of wagons over the site, dropping their loads on the experimental station in long sticks of trailing ordnance. "Our low group had most of its bombs fall in the immediate target area and had some hits directly," reported another bomber crew.

The P-51s were on every point of the compass, high up in the atmosphere, low at ground altitude, far ahead of the main body of the convoy of bombers, on all sides.

Mustangs dove on an airfield and a lead P-51 "shot up one He 111 on the ground which burst into flames and blew up as he flew over it. Some of the flying debris from the exploded 111 hit the wing of Lt. Alexander's plane and damaged it," one chronicle recalled. Two other P-51Bs piloted by a pair of lieutenants, Rader and Brucks, "made a pass on Barth airfield which proved to be enough, as the

20mm flak was very intense here. They each succeeded in getting strikes on two different Ju 88s, but the darn things would not burn and claims therefore are 1 Ju 88 damaged by Lt. Rader and 1 Ju 88 damaged by Lt. Brucks."

Mustangs unleashed their power all along the Baltic coast of northern Germany—Jutland, Pölitz, Sundhagen, Loissin and Koserow, cutting across the Danish frontier over the top secret works at Peenemünde and ancillary targets and installations in the surrounding area.

Now Lieutenant Colonel Robert H. Kaurin of the 381st Bomb Group saw above him some of the experimental jet fighters developed by the Nazis.

"A flight of Me 262 twin engine jets was about ten thousand feet above us and we were at twenty-eight thousand." The jets whooshed by with a maximum speed of 541 miles per hour from their Junkers Jumo 109-004B turbojets, capable of climbing 55 feet per second with a ceiling of almost 40,000 feet.

That day, the Mustangs also hit airplane-parts factories and struck at the great naval base at Kiel as well.

The 77th Fighter Squadron of the 20th Fighter Group now decided to go freelancing for ground attacks, targets of opportunity. The commander of the 77th, Captain Donald A. Reihmer, of Elmhurst, Illinois, presently found a German airdrome filled with planes; charging in in his P-51D, he led off on the tilt. As he swooped down, he noticed at least six German E-boats—similar to an American PT boat—and opened fire, destroying several and damaging the rest. Onward, in toward the aircraft, Heinkels and Junkers—right in Reihmer's sights. Then the pilot dove, sliced in . . . and opened fire. He picked out one of the planes—a Heinkel He 111—and screamed down on it and destroyed it. He successfully swung away and climbed fast—but his plane lurched, and in the next instant his engine cut out. He had been hit. Reihmer used the momentum of the Mustang, traveling at 350 miles per hour, to take him out over the open sea as he tried to restart his engine. No luck. He was all alone above the trackless wastes of the ocean. He prepared to bail out.

"I unhooked everything, wound my canopy open, pulled myself up and went over the [left] side, right foot first, and went head-first down over the left wing," he recalled later. He sailed down from his aircraft in the unbounded sky and pulled his rip cord with some difficulty, using both thumbs.

The parachute blossomed; he sailed down over the sea like a kite being taken in slowly and steadily. He was at 700 feet; he glided down, approached the water, straightened out his legs; then he plunged into the sea. He had a survival kit and a small inflatable one-man dinghy folded into a little bundle in his emergency gear; he fought to get out of his parachute harness. "The sea was rough and the wind was blowing into the open parachute and I was being dragged along the surface, meanwhile being inundated by every wave. It was hard to get a breath without swallowing the brackish water so [I] soon realized that it was either go under the surface to work out my escape or be drowned in the struggle." Reihmer was a good and strong swimmer, so he didn't mind diving beneath the surface to work on the parachute harness.

He got free, inflated the dinghy, crawled up the side and hauled himself in. Reihmer's entire world was reduced now to a tiny rubber dinghy adrift on a trackless sea.

His survival kit included French, Dutch and English currency, silk maps of those countries, some hard energy bars, hard candy and a pint can of fresh water.

The dinghy bobbed on the waves, alone on the 1,000-mile tract of empty ocean that surrounded him on all sides, a vacant, empty waste filled only with distance and the receding void of the surface. Reihmer did not know where he would head. He took stock: he figured his provisions would last him a couple of days if he rationed them properly. He rigged the little mast and sail of the dinghy.

He did not know how long he would be in the middle of the great alone. The sail blew him west and there was nothing at all on the wilderness of the sea and the empty range all around. "I was tossed and turned but the dinghy rode the waves very well. In fact, despite all . . . ups and downs . . . I don't remember shipping a lot of water. . . . As the night came on it got very cold. I was wearing a tanker jacket, tan with elastic cuffs and collar, and GI woolen trousers. The dinghy had a cover that came up to my neck and could be snapped shut. This not only provided a water shed but helped to keep out the wind."

At dawn he decided to head for the German shore. On the open sea, for hours under the hot sun of day, Reihmer kept what seemed an unending vigil for a mast, a speck, a smoke trail, anything approaching that might stop, might save him. "While sailing I prayed a lot, asking God to save my life." Captain Reihmer had been devout since his youth. "I quoted several verses from the Scripture to Him, which I took as promises to me personally. He honored that request. . . ." Reihmer paddled on, making for the shore of Jutland, occasionally taking a small swig of water.

"At about midnight, I had been sort of dozing in the dinghy with the sail between me and the extremely bright moonlight when I heard a very loud continuing noise." A ship. It was bearing straight down on him. "All I had time to do was to cry out, 'God help me!' and He did. . . ." He credits God for what happened next. The vessel stopped, lowered a boat; the crew picked him up. He climbed the ladder out of the boat and onto the deck of the ship.

"Are you English or American?" a crewman asked.

"American."

He was on the Swedish ship SS *Skarborg*, under the command of Captain Gunnar Robert Swanstrom, soaked and weathered and weak with gratitude in the middle of the wide North Sea. He was safe.

The *Skarborg* docked in Stockholm early in the morning on Tuesday, August 8, 1944. Captain Reihmer met with a U.S. military attaché from the American embassy. He was flown out by a B-24 to Scotland two weeks later.

Crossbow bombing was suspended with the last V-1 launch from France on September 3, 1944. It was resumed after the first V-2 attack in mid-September on launch sites in Holland, and continued until the end of the war. While Crossbow delayed the V-1s over England, overall the operation was not a success. The small rocket sites could be moved and repaired, and the large concrete sites had already been abandoned. And the cost to the RAF and the Eighth Air Force was high: Operation Crossbow would involve over 25,000 bombing sorties and drop 36,200 bombs before it was over. Four hundred bombers and 2,000 airmen were lost.

Beyond Crossbow, Hap Arnold had approved a deadly experiment to further cripple the V-weapons. On June 12, like Churchill, he had visited the Normandy beaches after the invasion. Accompanying Eisenhower aboard the destroyer USS *Thompson*, he had seen the majestic display of Allied ships massed in the Channel and the great system of artificial piers or Mulberries under construction to receive tons of supplies for the troops still going ashore.

Returning to England, he and other generals were awakened in the night by a crash and an explosion. It was one of the V-1s the Germans had begun to fire on London. Three hundred had been launched that night. When the all clear sounded, Arnold called for his car and drove to the site of the exploded bomb in a nearby village.

He had the remains of the V-1 sent back to Wright Field for analysis. And he approved a secret project code-named "Aphrodite." The idea was to pack old four-engine bombers—"Weary Willies"—with high explosives, fly them over V-weapon sites and then, as the two-man crew bailed out, use control ships (other aging bombers) to remotely guide them to the targets. On August 12, 1944, a B-24 Liberator crammed with explosives and piloted by a two-man crew was headed for Mimoyecques in the Pas-de-Calais in northern France—a huge underground gun emplacement from which a thousand rockets an hour were to be launched at London. Suddenly, the B-24 blew up in the air in a fireball, killing both crewmen. One of them, the jump pilot, was Lieutenant Joseph P. Kennedy, Jr., son of the former U.S. ambassador to London and the older brother of future president John F. Kennedy. The bodies of the two men were never recovered. Flying above Lieutenant Kennedy's plane was Colonel Elliott Roosevelt, head of the Eighth's reconnaissance unit and one of FDR's sons.

CHAPTER 24

Fighter Boys

The basic profile of the American fighter jockey was a combination of daredevil and cowboy, and many fighter pilots fit that profile. World War I ace Eddie Rickenbacker had been a race car driver in his youth; Colonel Hub Zemke had been a boxer; Lieutenant Ralph K. Hofer, a 4th Fighter Group ace, had been a boxer, too—a Golden Gloves trophy winner. Jimmy Doolittle had started out as a racing and test pilot, setting an airspeed record in 1932 of 296.287 miles per hour. Tommy Hitchcock had been a legend in polo.

Fighter pilots tended to be mavericks. Blakeslee had once been disciplined for having not one but two girls in his room. One pilot, in a macho storm, had downed a fifth of Scotch in one sitting to make a point. Hitchcock had flown under the Fifty-Ninth Street Bridge commuting to work in New York. Fighter pilots were of all kinds, but one trait was common to many of them—they did not much fear what they faced or feel much doubt or hesitation.

"I'm 19 years old. They gave me a plane with over 2,000 horse-power and six .50-caliber machine guns and they said, 'go have fun,'" recalled Joe Peterburs of the 20th Fighter Group, a pilot with six victories. "I had fun."

"You got the adrenaline probably was going a little more, jumped up a little bit," recalls then-Lieutenant Wallace E. Lowman, who was with Joe Peterburs in the 55th Fighter Squadron of the 20th Fighter Group, "but I don't think anybody was afraid. I won't say that's 100 percent; there was probably somebody out there that really didn't want to be there. None of us wanted to be there, but it was a job to do, and where do we go?"

William R. MacClarence, then a lieutenant and Mustang pilot in the 339th Fighter Group in Fowlmere who helped shoot down a German Me 262 jet, said of fighter pilots: "They had a swagger. You were very proud that you were a fighter pilot. And happy as hell you weren't a bomber pilot." He went on: "People ask me: Were you afraid? My answer is no. I think the reason was that we were so trained and so committed that you really didn't want to fuck up. . . . You know that book about *The Right Stuff* where [an astronaut] is sitting on top of the rocket and they ask him were you afraid? He says no. He says, 'My words up there were, please God, don't let me fuck up.' And that's the way we were."

The natures of fighter pilots varied. Lieutenant Ralph "Kid" Hofer, who like Blakeslee had enlisted with the Royal Canadian Air Force and joined the Eagle Squadrons, was a daredevil ace who flew by instinct, had unfaltering nerve and a tender, gentle side. With his baby face, laughing eyes, mop of hair and football jersey, Hofer had been an amateur boxer. The winner of a light-heavyweight Golden Gloves competition in 1940, he had given up boxing because he didn't like hurting people. Kid Hofer's constant companion at Debden was his enormous Alsatian dog, Duke. When Hofer was killed in action in mid-1944, the dog remained the mascot of the 4th Fighter Group.

Avy Clark, Tommy Hitchcock's nephew who became commanding officer of the 4th, was a high-society scion who smoked a pipe, wore tweeds and was a man of easygoing charm. John T. Godfrey, the leading ace of the 4th with 18 aerial victories, was an impulsive rebel who ran away from home twice as a boy and went into the Royal Canadian Air Force at the start of the war rather than go to college. Major Godfrey had a sweetheart, Lieutenant Charlotte Frederick, a nurse at a nearby hospital; they, too, had a dog, Lucky. Each time Godfrey took off on a mission, he would buzz Lieutenant Frederick's hospital and give three blasts on his throttle. Sometimes Lieutenant Frederick would come out and wave. Godfrey would buzz the hospital again on his return.

Many fighter pilots were ferocious spirits. They had something in common with the great barnstormers of the previous generation, flying stuntmen and circus performers like Charles Lindbergh and Bessie Coleman, a Black woman acrobatic pilot who had pulled off astonishing acts above a besotted crowd. They were risk-takers almost like the wild men of the next generation, men like Evel Knievel, who jumped fourteen Greyhound buses on a motorcycle in one lunge. But such tricks were planned carefully to eliminate as much danger as possible. To chase an enemy fighter plane at speeds over 400 miles an hour, risking your life against a foe who had every incentive to kill you first in a battle in which the next move could not be anticipated—that took a different kind of nerve.

These aviators not only took risks aloft. They often took them on the ground, too. Once shot down over enemy territory, they were clever and resourceful in staying alive and evading capture.

Captain Jack M. "Happy Jack" Ilfrey had three lives. Commanding officer of the 79th Fighter Squadron, 20th Fighter Group, Ilfrey once charged a German Messerschmitt Me 109 head-on when he had run out of ammunition, knocking the German out of the sky. Three weeks later Ilfrey was shot down over occupied France but avoided Germans and made his way to an American position in clothes supplied by sympathetic French peasants. On his return to base at King's Cliffe, he was ready to fight again.

In November 1944, when Ilfrey's wingman, Lieutenant Duane F. Kelso, was hit and lost power, Ilfrey watched him land on an abandoned strip in Holland. Defying ground fire and steep odds, Captain Ilfrey landed his Mustang next to Kelso's. He got out; Kelso climbed into the cockpit; Ilfrey stepped in and sat in Kelso's lap. With the two cramped and contorted like monkeys in a cage, Ilfrey opened the throttle; the plane lurched and labored down the runway, staggered to the end of the field, and lifted into the air with a shriek and a roar. Captain Ilfrey flew the plane out of Holland; the P-51 carried its two passengers away from danger and the city of Maastricht to Brussels, then toward home and a safe landing.

Captain Kendall "Swede" Carlson, from the little town of Red Bluff, California, never stopped fighting, even when he was shot down. Carlson, of the 336th Fighter Squadron, 4th Fighter Group, was strafing an airfield in Germany when he was hit. He managed to land safely behind enemy lines, then rolled back his cockpit canopy, stood up and, with his radio still working, began acting as a combat command coordinator, guiding and directing other planes from his outfit in a combined assault on the German base they had been attacking. He was eventually overrun and taken prisoner.

The "Ace of Aces," top Mustang ace Major George E. Preddy, Jr., was commander of the 328th Fighter Squadron, 352nd Fighter Group, and scored a boggling 26.5 kills in his P-51 *Cripes A' Mighty*.

Preddy shot down a record one-day total of six enemy aircraft in a single day and held the Air Medal with six Clusters, the Distinguished Service Cross and a Croix de Guerre from Belgium, among other decorations. His commander, General John C. Meyer, said the Greensboro, North Carolina, pilot was a man with a "core of steel in a largely sentimental soul." Preddy listed in his golden rules of life: "7. Always try to give the other man a boost. 8. Fight hardest when down and never give up." Preddy was killed in action on Christmas Day, 1944, when he was accidentally shot down by friendly fire. He had flown 143 combat missions in P-47 Thunderbolts and P-51 B and D Mustangs. He was a one-man fighter squadron.

Major James A. "Goody" Goodson, 23 years old, barely made it home.

Commanding officer of the 336th Fighter Squadron, 4th Fighter Group, the Mustang ace who flew on D-Day and on the Pölitz mission, Goodson was defending American bombers raiding a German U-boat base near Kiel when his plane was hit in the engine by enemy flak. The plane lurched in midair and there was a crump. Black oil hit the windscreen and the plane started losing power. Goodson feathered his suddenly stricken beast onward at 25,000 feet, holding her in the air, glad he was headed for home. He saw, though, that his engine temperature had skyrocketed and was at redline, burning up. Instantly, he remembered that a rich fuel mixture could bring down the temperature. Major Goodson reached for a primer and squirted a shot of straight gasoline into the engine; the engine temperature came down. Alone, with his plane shot up, and far from the Dover coast and home, Goodson began a torturous solo trial of grit and muscle, manually pumping gasoline into the engine. His hand began to bleed under his glove. All the way back to the coast of Britain, his arm never rested, feeding the over-heated engine. Hours later, Goodson made an emergency landing at an airfield in England near the North Sea; by then his altitude had dropped to 500 feet over the ocean beaches.

These pilots were brave men, but they were most of all aviators of virtuoso skill who could simply fly better with faster reflexes than most pilots in the air. They had different talents and different tricks, but all had mastery. Though each pilot had his own style, he was not always even aware of it. Some pilots flew on instinct. Hofer, an ace with 16 victories, would race out alone after enemy aircraft, attacking whatever he could find. But as Goodson recalled in his memoir: "The Kid was usually among those who reported victories. When asked how he did it, he would laugh and show his snake ring. 'I'm one of the lucky ones,' he would say.

"I tried to draw him into discussions on tactics, strategy, meth-ods of attack, deflection shooting, and all of the tricks of the trade that I was constantly discussing with the others. . . . The Kid would just laugh and say: 'Hell, if I worried about all that theoretical stuff, I'd never shoot anything down. I just go get 'im. I don't aim my guns at them. I aim *myself* at them.' What came to most of us after years of training, study, trial and error, came to pilots like Hofer perfectly naturally and intuitively."

Hofer was killed in action on July 2, 1944, over Budapest, during Operation Frantic, the Russia Shuttle. Weeks later the Germans confirmed through the Red Cross that he had been identified by his dog tags and buried in Hungary. He was called the last of the "screwball pilots"—a consummate gladiator who ranked 12th in the top 20 Mustang aces of all time.

They drew on skill, instinct, control, technique, but in the end fighter pilots were alone in battle, solo in the cockpit, with all the weight of the war resting in their hands. "BMac" MacClarence of the 339th, who helped shoot down the Me 262 jet, said of the pilot's métier: "It's most challenging. It's the most challenging endeavor I could imagine. You're there by yourself. Nobody to ask questions . . . Nobody would give you an answer," says MacClarence. "Just do it. . . . The whole mission is quick. You think it's quick. You end up with calluses on your ass. Your fanny gets numb after a while. You couldn't move around. Your legs were on the rudder. Time went fast. . . . Time went very fast."

This was the caliber of airman who flew the Mustang, good enough to beat what had been the greatest air force in the world in 1939. "You have the different types of aircraft and this sort of thing," says Joe Peterburs of the 10th Fighter Group. The design of the P-51 was of course game changing. "But in the end, it's the pilot. A pilot with no experience and a good aircraft isn't going to make it against an experienced pilot and a less capable aircraft. It's basically both. That's what makes the real difference."

These were the young men in helmets and goggles who, in the uncertain days of 1944, destroyed the German *Luftflotten* in the skies in six rapid months, the cavaliers in flight suits and Mae West life jackets behind the remarkable machine that had grown from Schmued's dream to Don Blakeslee's weapon of choice, the javelin they had passed on from hand to hand. These were the pilots behind the Mustang legend.

CHAPTER 25

A Last Throw of the Dice

By the fall of 1944, the Luftwaffe was reduced to a shadow presence; Blakeslee and his pilots kept waiting anxiously for the Germans to reappear in the skies. They kept thinking the fighters were being hoarded for some dramatic last throw of the dice. The reality was that although the Luftwaffe was just barely managing to replace its planes, it no longer had fuel to fly them. And it could not stem its loss of pilots. In the last phase of the war, the Nazis were sending out pilots so young, their life expectancy in combat averaged little more than a day.

In the last fervid months of resistance, Germany unleashed an astounding array of futuristic "ultra" weapons, advanced and deadly, a palpable danger that threatened to undo the apparent Allied victory in Europe. The V-1 and V-2 flying bombs and rockets developed at Peenemünde, which the 20th Fighter Group had raided in June, were hardly the only new threat.

The new *Vergeltungswaffen*, or retaliation weapons, were of all kinds:

naval, aerial and for land use. There were missiles, submarines, aircraft—a whole array of entirely new inventions that threatened to break open once more the whole field of battle.

At sea, advanced laser-fast "Walter" subs were developed, named after their inventor, Hellmuth Walter, with 100-ton batteries and electric engines. Underwater speed on a conventional Type VII U-boat was 7 to 8 knots; a Walter Type XXI sub could make 18 knots submerged, whooshing along at more than 20 miles an hour in the gloom of the deeps.

In the air, there was the Me 262 *Schwalbe*, which master aviation historian Walter Boyne called "arguably the most beautiful aircraft of the war." The German engineers and designers also delivered for Hitler the Arado AR 234 *Blitz*, the first operational jet bomber; the Messerschmitt Me 163 *Komet*, the first rocket-powered plane ever to fly and which, shaped like a sparrow, flew at an almost unbelievable 624 miles per hour; and the Gotha Go 229 stealth bomber, a flying-wing interceptor with a top speed of 702 miles per hour, coated with material that made it resistant to radar pickup.

All these pieces of wizardry were launched by the Germans in utter desperation. But they caused deep concern: Allied leaders feared that if the new "ultra" submarines and jets could be brought into battle swiftly enough, and in great enough numbers, they could reopen the conflict on the sea and in the air. The jet fighters especially struck fear among Army Air Forces brass: they were faster than conventional fighters and seemed poised to defeat them in the air.

The fighter groups of the Eighth Air Force would have to contend with these fiendish new arrivals—but flying against the German jets was one kind of combat Don Blakeslee would have to miss.

In September he had gone home on leave. After three years of combat without a break, he was reluctant to leave England, worried that once he set foot in America, the air force might not let him return. His buddy from the 4th, Deacon Hively, was already in New York, waiting for him at the Sherry-Netherland Hotel. Blakeslee flew home and, ensconced in a suite at the top of the Sherry-Netherland, faced reporters and photographers from the *New York Times.* "Hero of 500 Missions Flinches at Camera," the headline read. Obviously ill at ease, Blakeslee was as usual uncomfortable with his celebrity. "It's more fun facing a squadron of Jerries," he remarked.

Fairport Harbor, Ohio, Blakeslee's hometown, gave him a hero's welcome: a parade past 6,000 locals and a wristwatch from the Diamond Alkali Company, where he had worked in his teens. "You are the symbol," the general manager told him, "of everything we have worked and striven for since Pearl Harbor." Back in Fairport Harbor, without fanfare, he secretly married his hometown girl.

Blakeslee was allowed to return to Debden, but the pleasure of being back was short-lived. For three and a half years he had kept the huge number of hours he had flown a secret, only logging the time from crossing the Channel to landfall on the return. When he was borrowed to lead another group, he recorded no time at all. Asked, "When are you going to quit?" his answer was always: "Quit? Why, I'm just learning to fly."

But now the air force was onto him. And when Hub Zemke, a fellow ace and Eighth Air Force celebrity, bailed out over Germany, the jig was up. Blakeslee was called into headquarters to hear that he was grounded. He was too valuable to be risked any longer over Germany, too great a public relations asset and publicity hero. That fall, he would pass command of the 4th to Avy Clark, Tommy Hitchcock's nephew, and return stateside to a desk job.

The pilot who had been the most responsible for bringing the Merlin Mustang to the Eighth Air Force and to the battlefront, who had led the most celebrated fighter group of the European air war, was beached.

Meanwhile, Army Air Forces fliers had realized that the marvelous late-model Mustangs could sometimes outfly the German jets and outperform them. It appeared that the jets, because of their great speed, were unable to turn with the P-51s, giving the Mustangs an advantage in the air. Nervously seeking to confirm this, the Army Air Forces ordered the 4th's Mustangs to do some experimental combat flying against prototype RAF jet fighters.

Hanging around the group operations building on one of his last mornings at Debden, Blakeslee eavesdropped as the Mustang pilots were briefed for the practice jet mission. Then he stared wistfully through the window as the mechanics readied the Mustangs for departure. The floorboards of the building shook as the jet fighters and the Mustangs taxied and took off.

With a few days left at Debden, Blakeslee flew for the last time, a series of sentimental sorties over the green English countryside to places familiar from his four years of war. He landed at Biggin Hill, the RAF station where he had flown for the British with the Eagle Squadrons after the Battle of Britain. Few of the RAF officers he had flown with were still there. In fact, of the 150 pilots he had started with in 1941, only three were left. On his last day at Debden, Blakeslee made a final flight in his beloved Mustang, insignia WD-C, to a nearby airfield. Lost in thought, the pilot who had flown 1,000 combat hours failed to let his wheels down. His Mustang crash-landed. Blakeslee was unharmed, but his war was over.

He returned to Debden, packed his gear into his duffel bag and was gone.

In the fall of 1944, the P-51s went after the jets with a vengeance. On the first day of November 1944, a 20th Fighter Group Mustang shot down an Me 262 jet. The next day the 20th's commander, Colonel

Robert P. Montgomery, led the unit in a romp—shooting down two Me 109s and a Focke-Wulf Fw 190 in a head-on charge. Three kills were chalked up by Lieutenants Harley Brown and Ernest Fiebelkorn; two Me 109s were shot down as they were landing at their base. In all, the group accounted for 28 enemy aircraft that day.

On November 8, on a mission to Merseburg, two P-51s charged another Me 262, which went spiraling down and into the ground without either of the Mustangs firing a shot. A few days later, Lieutenant Brown polished off a locomotive in a strafing run. In December several more Me 109s were bagged by the outfit.

Avy Clark had taken over command of the 4th Fighter Group on September 1. On November 2, the 4th's Mustangs met their new adversaries: Me 163 *Komet* rocket fighters, recently launched by the Nazis. Four out of five "Fleas" were cut down by Major Louis "Red" Norley and three other pilots.

Two weeks later, on November 18, the group found jets again, this time parked on an airfield. Swooping down, they fired on and plastered 12 Me 262 *Schwalbe*s. One did try to take off, but the men of the 4th managed to blow the plane away before it could gather enough speed to be airborne.

The Mustang was chewing up every ghoul the Germans had. Messerschmitt, Focke-Wulf, Junkers, Dornier, Heinkel, jet or rocket—it didn't matter to the P-51s. It was the same everywhere, from Cologne, to Berlin, to Wilhelmshaven in the north and Munich in the south: nothing could keep up with the Mustangs, even the jets and rocket planes. The sky was scrawled over with the signature of Edgar Schmued's P-51D.

Later in the war, Colonel Joe Peterburs of the 20th would tangle with one of these modern technological marvels, an Me 262 *Schwalbe* jet. He recounts: "The mission was one of the largest during the war—over Oranienburg—there were over 1,000 B-17s; and about 900 to 1,000 fighters escorting them." Suddenly they were in a fight. A covey of seven Me 262 jets came racing through the ranks and corkscrewed, porpoised and dove through the Flying Fortresses. Peterburs saw one of them blow up a B-17, then destroy another. Now he visually locked onto one of the jet fighters twisting through his flotilla and squeezed off .50-caliber rounds, pouring fire into the plunging aircraft.

"I hit him. He rolled over and started down to the deck and I was chasing him down," recalled Peterburs. "I started losing my speed advantage and we got down to about four to five thousand feet and there was a cloud layer there and he went into the clouds and I said 'I've lost him.'"

Not until many years later did he hear anything further of the mysterious jet fighter he had fired on. He learned the Me 262 had been piloted by Oberleutnant Walter Schuck, one of Germany's highest-scoring aces, with 206 air victories in over 500 missions flown since joining the Luftwaffe in 1937. Schuck had flown an Me 109 in the Arctic Ocean fighter squadrons, "becoming the Russian air force's feared enemy after he shot down 12 aircraft in one day."

That day Joe went on to strafe an airfield. He blasted away four to five aircraft on the ground and set fire to a hangar. Then "I felt a thud . . . and then I felt another thud and I started getting oil streaming from the lead. I made it up to 10,000 feet."

He had been struck and his plane was going down.

He descended to 1,000 feet now and knew he must bail out. Just then a Focke-Wulf Fw 190 pounced upon him and fired rockets, which missed. "I'm down to around 400 feet and I get off on the wing. I have to bail out on the wrong side of the aircraft, because there's a fire on the right; and I bail out, I let go and I hit my right knee on the horizontal stabilizer; I pull the rip cord, the chute opens; I hit the ground. And hard."

Joe was wounded and alone, shot down in German farmland.

Soon some farmers came upon him, then German Army soldiers. He was taken to a town jail, and then to a prison camp. Before long, he escaped, joined an Allied Russian tank unit until he came upon a U.S. Army patrol. They welcomed him in. He was back among his own, eating K rations again. From there, by twists, turns and sheer force of will, Joe made it to the French seaport of Le Havre, next to an air base, and headed for England in a B-17.

Fifty-three years after his remarkable experience with the jet, Peterburs received a letter from Germany. The writer had been a 13-year-old hiding in a ditch, watching the air battle up in the clouds. The memory haunted him, and years later he excavated the field where Peterburs went down. He unearthed parts of the plane—its guns, its engine, its prop—as well as its serial number. This led him to Peterburs's identity and, after a tireless search of Luftwaffe records, to the identity of the pilot of the German jet Peterburs had shot down.

Schuck, after shooting down four B-17s on that day—April 10, 1945—was hit by fire from Peterburs's P-51. Bailing out, Schuck sprained both ankles on impact and was still recovering when the war ended a month later. It was the last flight of his futuristic jet.

The boy who had stared up at the jet and the P-51 dueling in the clouds brought the two pilots together. They remained friends until Walter Schuck's death in 2015.

CHAPTER 26

Twilight for the Reich

Then the battle that had gathered and marshaled through the fall as both armies collected their strength and supplies finally broke out into the last stand. Early on the morning of December 16, 1944, Hitler sent his massed Wehrmacht legions hurtling across the line of demarcation against the Allied armies in a last, desperate effort to turn back the drive that had narrowed his kingdom down to the original lands of Germany. Eight armored divisions and 200,000 troops took the Allies by surprise in the Ardennes forests at the weakest point in the American defenses and breached their line; the Fifth Panzer Army penetrated to points within 20 miles of the Meuse River, at the towns of Givet, Dinant and Bastogne. The weather was frigid, wet and misty, the biting winter cold and fog persisting for eight straight days. No aircraft could fly, and the Eighth Air Force could give no cover or backup to the GIs caught in the snows.

"A bitter, dense, freezing fog settled across northwest Europe.

German jet reconnaissance planes reported to the German High Command at Zossen that all of East Anglia was completely socked in by fog," recalled Frank H. Lewis, an airman with the 491st Bomb Group. "The big bombers of the Eighth Air Force were grounded. These circumstances were ideally suited to the Germans' needs. They had been waiting and planning for just such weather."

"Fog, night and snow," Hitler predicted, would be the undoing of the Allies. And for days, as dense cloud cover kept the Eighth on the ground, heavy snow in the untracked forest allowed Panzer troops to drive the Allied lines far back, creating a great pocket or bulge in the American ranks, from which came the name "Battle of the Bulge." Nine hundred Wehrmacht tanks and massive Luftwaffe squadrons spearheaded the attack; by December 24, the German thrust had reached 65 miles into Allied lines along a 20-mile front.

Intense fighting continued for days. The Americans poured reinforcements into the Ardennes. But for more than a week of ice and wind, the entire Allied effort sagged.

Then, as Lewis recounts, "On December 23, as the Germans moved into France and Belgium, a young Lieutenant in the weather room at the Eighth Air Force Headquarters in London was looking at the weather charts and made a fantastic weather prediction. If the pressure ridges reacted as he thought they might, there was a better than even chance that a high-pressure ridge would develop over the target area on the 24th, allowing just enough time for the bombers to get in to the target and return to their bases."

The bombers could fly. The Eighth could take wing. "The field orders came clattering over the teletype at the 491st Bomb Group in the small hours of Christmas Eve. A maximum effort with every available plane was demanded." The weather had broken. The skies were clear. Pilots, gunners, navigators raced for their planes.

The "Mighty Eighth" took off to defend its countrymen in the infantry.

More than 2,000 bombers, shepherded by Mustangs, flew to raid German positions that morning. The Fifteenth Air Force put up another 1,000 B-17s and B-24s over southern Germany. With 1,000 fighter planes shielding the bombers, a total of 4,000 aircraft in a magisterial array closed the attack on the Germans that Christmas Eve. The bomber stream was over 400 miles long; airplanes were just leaving England as the lead bombers were arriving over German targets.

Throughout the Battle of the Bulge until it ended on January 25, Mustangs of the 20th and other units ruthlessly pounded German positions, their fervor unleashed by that Christmas Eve break in the weather. Over 70 North American Mustangs of the 20th joined the Christmas mission in the fog and cold. From December 25, the 20th Fighter Group flew almost daily missions over Wehrmacht positions, covering American troops below and making sure no reinforcements could move up. An average of 54 fighter planes took part in each of these final December missions.

In one high-tension attack, Major Richard P. Gatterdam of Ross, California, ran into a gaggle of Fw 190s and Me 109s and hit them directly. Almost immediately another flight of Nazi raiders appeared; other Mustangs went in for the bounce. Soaring, speeding, veering, the 20th fighters savaged the bandits, Gatterdam accounting for two German planes. On his last bag, Major Gatterdam watched the pilot bailing out of his plane with one foot on his wing and one in the cockpit. The wing then came right off the German's fuselage, and both wreck and pilot plunged to the ground below. The 79th Fighter Squadron of the 20th claimed a grand total of six German aircraft destroyed and three damaged in that one action. Magdeburg, Stendal and Brunswick were also hit the same day.

On that Christmas bombing run, a moment of intense drama, every B-17 played the role of Santa's sled, delivering presents of TNT to the Germans. Lewis and his plane reached the IP (initial point) at 1:42 p.m. Their target: the Wittlich Bridge. "The outside temperature . . . was rather warm for Christmas over Germany. . . . The snow-covered fields and forests below looked quiet and peaceful. The Christmas-card-like scene belied the death and destruction below."

Then, at two minutes past 2:00 p.m., the drop: "I watched the bombs all the way down," Lewis said. "The squadron bombs had a good pattern, starting at the west end of the bridge and walking all the way across to the other end, knocking both spans of the highway and the railroad bridges into the canal.

"I'll bet it was one of the greatest Christmas gifts our ground troops will ever get." The air crews would be home in time for a turkey dinner; at 4:42 p.m., all aircraft were secure back on base in England.

In Washington, FDR, who had been jubilant from his election to a fourth term as president, was in deteriorating health by Christmas. He looked exhausted, his blood pressure was up since the campaign and, despite being fed eggnog, he had lost weight. Three weeks at Warm Springs in December had not helped. And the events in the Ardennes caused him great anxiety. His Christmas Eve message, broadcast on the radio and delivered as the Army Air Forces were already over France, bringing relief to Allied troops, was "more full of hope than cheer." Presiding over Christmas dinner with his grandchildren, he looked like "a gaunt old patriarch." His secretary remembered the president looked weary. By April he would be dead, four months before the war he had led came to an end with the Japanese surrender aboard the USS *Missouri* in Tokyo Bay.

The Battle of the Bulge spanned Christmas and New Year's that winter: the air cavalry charge of the Eighth Air Force was critical in turning the tide of the battle. Attacking through December 25, the air offensive froze the Germans in place. At last the Wehrmacht's lines broke, with tank crews running out of gas. Their advancing armies slowed and came to a halt: they had only 36 hours of fuel left to drive their motorized units. The armies abandoned their weapons and walked home, the retreating Germans blowing bridges over the Rhine behind them. By January 25, the Nazi front in the Ardennes had collapsed. Final American casualties on the ground would be 75,000; the Germans would lose 120,000 in the snow-covered forests of Belgium and Luxembourg.

Now on the far side of Germany, the Russians closed from the east, from the border with Poland, Czechoslovakia and Austria; they brought up their own massive supplies, strengthened by the arrival of innumerable American trucks for motorizing their infantry. In December, forbidding reports had come to the new German chief of staff, General Heinz Guderian, the hero of the French collapse in 1940, that 225 Soviet infantry divisions and 22 armored corps had assembled on the Eastern Front between the northern Baltic Sea and the Carpathian Mountains. Hitler had jeered at the threat, declaring, "It's the biggest imposture since Genghis Khan! Who is responsible for producing all this rubbish?" The Führer, in denial, had refused to halt the Ardennes assault or transfer forces from the west or the Baltic Sea to counter the Russians.

On January 12, with the Battle of the Bulge still in progress, the Soviet offensive launched out. Three Russian armies under Generals Zhukov, Konev and Rokossovsky rapidly broke a huge gap in the Wehrmacht's front 200 miles wide. On January 17, Warsaw was captured by Zhukov; two days later spearheads of Zhukov's force drove into Łódź. On the nineteenth, Konev's troops reached the prewar Silesian frontier of the Reich. The Russian advance through the breach—100 miles deep and 400 miles wide—was a mortal blow. Hitler's empire was beginning to crumble; the Reich was disintegrating. The Führer could not at the same time guard his front door from the Americans and his back door from the Soviets.

All three Russian formations moved ahead fast; on January 31, 1945, the Russians were 40 miles from Berlin. Now the two jaws of the Allied maw—the Russian on the one hand, and the American, British and Canadian on the other—began to shut. The closing Russian advance convinced Hitler to shift his newly conscripted home guard to the eastern lines. This tip of the seesaw to the east released pressure in Central Europe, opening the way for the advance of Allied troops and for a crossing of the Rhine River from the west.

The Allies now slammed into the German interior. The advance was deliberate and swift, meeting little resistance. Patton, Bradley and Montgomery, fighting their way into the German homeland, raced through the last distances of the western frontier and the roads to Hamburg, Düsseldorf, Frankfurt and the capital.

American airpower fought above and alongside the ground troops. On February 3, 1945, the Eighth hit Berlin again, destroying the huge Templehof marshaling yard as well as the headquarters of Hitler's government—the Reichschancellery, the Ministries of Air and Propaganda, the Gestapo headquarters. In the process, some 25,000 civilian Berliners were killed and 120,000 left homeless.

A week later on February 13, preceded by the Lancasters of Harris's RAF, and again on March 2, the Eighth Air Force attacked the marshaling yards at Dresden. As in many towns and cities throughout the Reich in the last months of the war, Eighth Air Force bombs intended for the railways spilled over into residential neighborhoods. Incendiary bombs caused a firestorm with hurricane-force winds that swept the city. In the words of Kurt Vonnegut, the American novelist, then a GI and a prisoner at Dresden during the attack, "the loveliest city that most of the Americans had ever seen . . . was like the moon" after the raid. It was "nothing but minerals." Between 35,000 and 40,000 Dresdeners died.

The Eighth Air Force had now spread American wingspan from the Baltic to the Rhine, Poland and Austria. Unopposed, they did not back off or ease up. It was a long, open, free run.

With the bombing barrage unfolding in the frigid winter, the Mustangs continued to break open the skies and fight through the last Nazi defenders: jets, rocket planes, missiles. On January 21, a pride of 20th Fighter Group P-51s warded off 14 Me 262 jets attacking a photo-reconnaissance mission. In February, they hit transportation and oil facilities. On February 9, on an escort mission to Litzendorf, the 77th Fighter Squadron of the 20th, led by Major Merle J. Gilbertson, joined in a raiding party. Major Gilbertson got three enemy birds, but one blew up in his flight path, shattering his cockpit canopy and wounding him. Not far away, near Esperstedt, Captain Frederick Larsen tagged an Fw 190. And other pilots of the 20th, too, carried on the rampage. Within 20 minutes they destroyed 39 Boche planes, with Captains Lowell E. Einhaus and Charles H. Cole accounting for 12 between them. On February 14, Captain Jake Brown of the 55th Fighter Squadron tangled with three Me 262s, damaging one. The rally went on. On February 19, the 20th made strafing runs in the Magdeburg area; ten locomotives were destroyed. On the twentieth, the group got another Me 109, then another four enemy aircraft in the air and three dispatched on the ground.

The "Fourth But First" bagged two Focke-Wulfs on February

20; on February 22 they caught eight fighters on the ground at Halberstadt Airdrome and set all eight afire. On February 25, they destroyed 13 more German fighters, including six in the air. The next day, after escorting bombers to Halle, the crack unit came upon a large number of pursuit planes on the ground at Weimar Airdrome and set 43 of them afire, leaving flaming wreckage. Everywhere, against even the most advanced Nazi machines on the ground and in the sky, the Mustangs dominated.

By the last week of March, the main strength of the Allied ground forces was collected at the Rhine River. Now the American and British forces unleashed their final drive. The great push across the Rhine began as four American armies prepared to cross the river on barges and pontoons. Patton's Third Army reached the Rhine at Koblenz on March 10. Still farther downriver, the 9th Army of General William H. "Big Bill" Simpson, the calm and dependable general who had been Eisenhower's classmate at the Army Staff College in Fort Leavenworth, Kansas, reached the Rhine at Düsseldorf. All the armies were now ordered to hold up for British Field Marshal Bernard Law Montgomery's decisive lunge. The aloof, haughty Montgomery was a prima donna, and he was determined to be the first across. But the Americans cleared the west bank, and on the night of March 22–23, Patton crossed the stream at Oppenheim between Mainz and Mannheim on pontoons at about 10:00 p.m. almost unopposed. It had proved astoundingly easy. On the night of March 23–24, Montgomery attacked across the Rhine with fully 25 divisions advancing across a 30-mile stretch of river near Wesel, a barrage from 3,000 guns and flights of bombers overhead softening up the opposition below.

During the night of the twenty-third at Debden, RAF planes could be heard overhead en route to bomb the Rhine defenses. As the sun came up, hundreds of gliders were being towed at low altitude toward Germany: Operation Varsity was on, the largest airborne assault in history, landing 16,000 paratroopers on the far, eastern side of the Rhine. The heavies massed in a cloudless sky, sparkling in the sun as 70 Mustangs of the 4th rose up to join them. Their orders were to stop everything that moved, in the air and on the ground.

All through Varsity, the 4th provided top cover, patrolling a 30-square-mile area around Osnabrück. They flew two missions that day, providing continuous impenetrable air cover for the Allied armies below from eight in the morning until six thirty at night. No German aircraft penetrated the 4th's aluminum awning.

Elsewhere, the 20th Fighter Group was assigned to sweep a sector around Münster, preventing any ground reinforcement by the Wehrmacht, as well as knocking off any air attacks. The end was in sight.

In the air, the Germans would fight to the last in a fever, never relenting or relaxing the ardor of their cause. In April, in two brutal air battles, Mustangs would eliminate the last resistance of the Luftwaffe. On April 7, 120 German student pilots climbed into the clouds to the strains of patriotic music blaring through their headphones to join the elite JG 7 (*Jagdgeschwader*) jet unit confronting a force of 1,200 Allied bombers and 800 fighters. Three-fourths of them were lost. Three days later, the P-51 of Joe Peterburs of the 20th was among the many that sortied from England to attack Luftwaffe jet fighter bases; more than 300 German aircraft were destroyed. These losses of jet aircraft would be the final blow to the Luftwaffe, and the air defense of central Germany was abandoned.

In Eighth Air Force lore, April 10, the day Peterburs shot down Walter Schuck, the German ace who became his friend, would be known as the day of the "great jet massacre."

The flame flickered down to its last as the Allied armies swept ahead, and the final, decisive drive for Berlin itself went forward. They closed in a narrowing arc, turning in from the Canadian First Army in Holland, in the north; to the American 9th and 1st Armies in Magdeburg and Leipzig, in the center; to the Free French 1st and 7th Armies on the border of Austria. On March 6, the Allies took Cologne. There they stopped.

On April 2, one month after capturing the city, American troops celebrated mass in her soaring, bomb-damaged cathedral, which despite 14 hits from aerial bombs remained standing in a city otherwise completely flattened. On that Monday, they paused and prayed. Then the advance swept on.

The Allied armies sped on to the Reichstag, drawing in the net. Like a trawl collecting and closing, the armies of General Eisenhower—American, British, Canadian and French—all converged on Berlin. They moved now almost entirely freely, after six years of grinding battle, all now speeding forward in triumph. They swept on, through Haldensleben, Goslar, Salzgitter, Uelzen and Soltau; on April 11, they reached the Elbe River 60 miles from Berlin, where Adolf Hitler had held out, hoping for some extraordinary turn of events that would save him.

Nazi Germany, darker than fable, was ruined. The serpent had no more head, or body, or tentacles. Hitler, alone and without hope, married his mistress, Eva Braun, after midnight on the night of April 28–29. The following day, April 30, the two committed suicide together. Grand Admiral Karl Dönitz of the German Navy, according to Hitler's wish, was installed as the last Führer and turned to saving as many troops fleeing from the Russians as he could.

The Soviet Army moved off its positions on the Neisse River and headed into Berlin; within a week Soviet soldiers were entering the suburbs, the forces of Zhukov and Konev completely surrounding the capital. After six relentless years, Soviet and American forces linked up at the Elbe. The war was over. Germany was in ruins. The Thousand Year Reich was an underworld vacant and destroyed, its lands a tract of rubble and cinders, its armies in tatters, its people without command, their geography a smoking, empty wasteland. Nazi Germany had tumbled to its end. It was a devastation rained down by the bomber sky rams that had pounded the land since before D-Day, an assault made possible by the P-51 Mustang, by the terrible swift sword that had come into Edgar Schmued's hand five years before.

And the Mustangs parked for good. The last mission of the 20th Fighter Group was Mission 312 flown that day, Wednesday, April 25, 1945. Only two pilots failed to return, Lieutenant Robert H. King, who parachuted to safety, and Lieutenant Lynn P. Bishop, who lost all his coolant fluid and bailed out near Düsseldorf. He was last seen safe on the ground in the company of friendly forces who waved to his squadron overhead that he was okay. They had watched his chute billow, spread and float him down, narrowly missing a chimney in the heartland of the Ruhr.

On that day Major Merle J. Gilbertson of the 77th Fighter Squadron, 20th Fighter Group, completed his hundredth mission.

Lieutenant Bradford L. Collins, of Laguna Beach, California, completed his first mission with the 20th, just as the curtain dropped on the show.

The Mustangs stopped, pulled up, gathered and sat on the runways, weary birds after a long flight, gleaming mirrorlike, silvery in the sun, angled back on their landing wheels like gulls, after their almost unbelievable, last-act, lightning-fast 18-month victory had brought their bombers, their country and their allies to peace.

The Bells Toll

The unconditional surrender of Germany was signed on May 7, 1945, in a schoolroom in Reims, France, at 2:41 in the morning. The schoolroom, which had been the Germans' supreme head-quarters and then Eisenhower's temporary command center since February, was hung with maps of Europe that he had pored over daily as he deployed his troops in the last months of the war. Now General Alfred Jodl, signing for Germany, was seated facing the maps—forced to confront all the territory the Reich had lost. The surrender was to take effect at midnight. All over England, village churches were permitted to ring their bells again for the first time in the six years since the war had begun.

In London that night, a thunderstorm broke with, as Mollie Panter-Downes wrote in her diary, "an imitation of a blitz so real-istic that many Londoners started awake and reached blurrily for the bedside torch." The following morning of May 8, designated as V-E Day, Panter-Downes recounted that "thousands of King

George's subjects wedged themselves in front of [Buckingham] Palace throughout the day . . ." After lunching at the palace with the king, Churchill joined him on the balcony with the royal family. "When the crowd saw Churchill," Panter-Downes wrote, "there was a deep, full throated, almost reverent roar."

The Eighth Air Force, a vast population of 200,000 people in the midst of the English countryside, stood down at its airfields from Leiston to Podington, from Boreham to Sculthorpe. Some personnel would remain in Europe; some packed up to go home; some prepared to transfer to the Pacific for the final assault on Japan.

In the aftermath of conflict, to give ground personnel an eyewitness view of the destruction their planes had wrought over almost three years, a program of observation flights over Germany was begun by the Army Air Forces. Some 10,000 men and women rode on these "trolley runs" to see the devastating damage to bridges, railways, factories, munitions works and oil facilities their efforts had helped to launch.

"We formed up in the B-17s, perhaps a dozen people in each plane, and we flew across the English Channel," Tom Stanback recalls. "We got to Germany, down over the Ruhr Valley, would drop down, low, just 700 feet, right down over the ruins. My God. It was dreadful. Just rubble. Everything was just wrecked. . . . Everything was torn to pieces. The buildings were bombed out, the factories. It was just a mess. We left early in the morning and didn't come back until suppertime."

The previous month, a small air force team flying over the cities bombed by the Americans and the British recorded the destruction the Allies had wrought: "Kassel . . . just miles of rust staring to the sky . . . Leuna . . . an enormous desert of iron skeletons . . .

Magdeburg . . . another ghost city . . . Cologne . . . indescribable. One gets a feeling of horror; nothing, no thing is left." One journalist on the ground said, "The cities they had passed through were not living communities; they were wounds in the earth."

While these trips unfolded, other B-17 units flew relief missions, bringing out thousands of prisoners of war and displaced persons from Austria to France, many sitting on lumber platforms hastily rigged across the bomb bays. From accounts of returning POWs and examination of destroyed German aircraft on enemy airfields, the claims of kills by fighter pilots could be confirmed. Blakeslee's 4th Fighter Group was the highest-scoring unit in the Eighth, with 1,016 German planes shot down or destroyed on the ground.

By the war's end, 15,000 P-51As, Bs, Cs and Ds had been milled out of North American's plants in Los Angeles and Dallas and shipped to the European Theater of Operations. They had flown more than 200,000 missions and over a million combat hours. P-51s had destroyed almost 5,000 German planes in the air and more than 4,000 on the ground. This was almost half the enemy aircraft destroyed in Europe by American fighters. The Mustangs had cleared the European skies.

These missions came at a high price: the Eighth Air Force lost 26,000 air crew. Only submarine crews in the Pacific suffered a higher fatality rate. Of the 26,000 Air Force casualties, some 2,100 were fighter pilots. British losses were far higher: some 3,700 were lost in fighters and close to 56,000 in bombers. The Germans suffered 70,000 casualties among air crews.

Mustangs went on to scour the skies in the Pacific. As the war in Europe wound down, the first P-51s were deployed to the Far East. In the spring of 1945, Mustangs were stationed on the captured island of Iwo Jima to escort B-29 Superfortress bombers on missions against the Japanese mainland. Large tracts of Japan's six largest cities were destroyed, and as a result, many Japanese realized the country could no longer be defended.

By the time of the Korean War, the jet age had arrived and transformed combat. Even in the age of Mach 1 and throw weight, the Mustang proved a superb warrior in Korea against the Russian-built MiG-15 jets. After the Korean War, the Merlin Mustang became one of the most popular racing planes in America: one P-51 attained an airspeed of over 500 miles per hour. P-51Ds are still restored and flown today.

The U.S. Army Air Forces became the U.S. Air Force, a separate, independent branch of the military, on July 26, 1947. General Hap Arnold, who had shepherded it into being, became a five-star general of the Air Force in 1949, the first person ever to hold the rank. After suffering four heart attacks between 1943 and 1945 under the extreme pressure of running the air war, he had returned to his ranch in Sonoma, California. He suffered a fifth heart attack and died in Sonoma in 1950. He never fully acknowledged or elaborated on his mistakes and misplaced assessments of the P-51 Mustang, even in his memoirs.

After V-E Day, Carl Spaatz transferred to the Pacific. In command of the U.S. Strategic Air Forces, he directed the atomic bombing of Hiroshima and Nagasaki. He was the only general present at all three surrenders: in Reims, in Berlin and aboard the battleship *Missouri* in Tokyo Bay.

General Ira Eaker, after his reassignment as air commander in chief of Mediterranean Allied Air Forces in January 1944, personally led the first mission of Operation Frantic in July 1944, flying in a B-17 called *Yankee Doodle II* to the Soviet base at Poltava. Almost 40 years after his retirement, Congress passed special legislation making him a four-star general.

During the war, Robert Lovett oversaw the massive expansion

of the Army Air Forces. In his long and distinguished career he served three U.S. presidents: Truman as Secretary of Defense, Eisenhower and JFK.

Dutch Kindelberger headed North American Aviation through its glory years, from 1934 until 1960. He would lead the small upstart company to produce 42,000 airplanes by the end of World War II, more military aircraft than any other airplane maker in U.S. history. By the 1960s, North American's Rocketdyne division was producing the engines for the Redstone, Jupiter, Thor, Delta and Atlas missiles and for the NASA Saturn family of launch missiles, which propelled the Apollo missions into space and all the way to the moon.

Edgar Schmued continued developing and pushing the form of the Mustang until the end of the war. The experimental lightweight P-51F/G became the P-51H; 1,000 were ordered by the Army Air Forces in June 1944. The Mustang master remained at North American until 1952, designing the F-86 Sabrejet—the fabled transonic fighter that saw some of the earliest jet-to-jet battles in history in the Korean War—and its successor, the veteran F-100 Super Sabre, which flew in Vietnam. He continued to consult on aircraft design until his death in 1985.

The *New York Times* front-page story on Tommy Hitchcock's death called his chronicle "one of the most gallant and one of the most spectacular careers in modern American life." Writing to Hitchcock's widow from the embassy in London in April 1944, Ambassador Gil Winant called Hitchcock's crusade "tangible evidence of Tommy's contribution to victory. Without it," he wrote her, "we would not be winning the war over Germany."

After the war was over, Gil Winant returned to New Hampshire exhausted and beset by financial woes. A love affair with Churchill's daughter Sarah had ended badly, and Roosevelt's death had robbed him of a close friend and mentor. Deeply depressed, he committed suicide in November 1947, on the same day his memoir of wartime London was published.

Don Blakeslee flew more missions against the Luftwaffe than any other American pilot. Desperate to stay in the war, he had a buddy falsify his flight log to reflect many fewer hours than he had actually flown. By war's end, Blakeslee had chalked up 500 missions and 1,000 combat hours, the most missions logged by any American ace in Europe or the Pacific. He remained in the air force until his retirement, flying in both Korea and Vietnam. He died in Florida at age 90.

Wilbur Wright died in 1912 of typhoid fever at the age of 45. But Orville lived on in the old family house in Dayton, Ohio, not far from Wright Field, receiving accolades from aviation pioneers all over the world. One day after the war was over, he found himself on a commercial flight. As the plane landed, the flight attendant, oblivious to Orville's identity, asked him if he had had a good flight. "Yup," he said. "I've had a good flight."

"Was it your first flight?" she asked.

"No," he said with a twinkle in his eye, "but that was a good flight, too."

Lady Lucy Houston, who had funded Britain's 1931 air entry for the Schneider Trophy, went on to support the 1933 Houston–Mount Everest Flight Expedition, on which aircraft flew over the summit of Mount Everest for the first time. In 1936, anguished by the abdication of Edward VIII to marry the divorcée Wallis Simpson, she stopped eating and died of a heart attack at age 79.

Charles Rolls, the business half of Rolls-Royce, had met the Wright brothers in 1909 and bought one of their early flyers. In

1910 he became the first man to make a nonstop double crossing of the English Channel. That same year he would be the first Englishman to be killed in a powered aircraft when the tail of his Wright Flyer broke off during an air show at Bournemouth. He was 32.

Knighted in 1930, Sir Henry Royce became a semi-invalid from chronic overwork. Continuing doggedly from his homes in Sussex and the South of France to work on the new engine design that would become the Merlin, he made the last drawing from his deathbed just hours before he died in 1933 at age 70.

And Ronnie Harker retired to New Zealand. He kept flying Mustangs. Both the New Zealand and Australian air forces had acquired large fleets of P-51s after the war, and one was always available to Harker. He would strap himself in, taxi out to the runway and take flight, lifting into the sky, rolling, banking, diving, relishing in his hands the plane he had brought to life and the epiphany that had led him to envision a Merlin Mustang so many decades before in the broad blue skies over Cambridge. Until 1997, two years before his death at age 90, Harker continued to fly the plane he had taken from obscurity to victory, from orphan to heroine.

In the summer of 1944, as the Allied armies advanced into France's interior, Antoine de Saint-Exupéry, who had so compellingly chronicled the intrepid wood and wire airmail pilots of early aviation, disappeared from the sky. That July he took off from Corsica on a reconnaissance flight for the Free French and never returned. It is believed he was shot down by German enemy aircraft. Born at the turn of the century within months of Edgar Schmued and Tommy Hitchcock, he left behind him *Pilote de Guerre (Flight to Arras)*, about his military pilot days, and many other works that are lasting testaments to the earliest age of manned flight.

APPENDIX A

Dates on Which Fighter Groups Acquired P-51s

Date Received	Unit	Base Location
December 1943	354th Fighter Group	Boxted
January 1944	78th	Duxford
	358th	Leiston
February 1944	4th	Debden
March 1944	355th	Steeple Morden
March–April 1944	352nd	Bodney
April 1944	59th	East Wretham
	339th	Fowlmere
May 1944	361st	Bottisham
D-Day, June 6, 1944		
July 1944	20th	King's Cliffe
	364th	Honington
	55th	Wormingford
September 1944	361st	Little Walden
	479th	Wattisham
[Retained P-47 Thunderbolt:	56th	Horsham St. Faith]

Bowman, Martin W. *8th Air Force USAAF 1943–45: P-38 Lightning, P-47 Thunderbolt and P-51 Mustang Squadrons in East Anglia, Cambridgeshire and Northamptonshire.* Barnsley, U.K.: Pen and Sword Aviation, 2009.

APPENDIX B

Mustang Production: Mustangs Accepted by the U.S. Army Air Forces

1943	Jan	Feb	Mar	Apr	May	June	July	Aug	Sept	Oct	Nov	Dec
TOTAL 1,710	–	–	70	121	121	20	91	175	201	284	295	332

1944	Jan	Feb	Mar	Apr	May	June	July	Aug	Sept	Oct	Nov	Dec
TOTAL 6,904	370	380	482	407	580	579	569	700	663	763	709	702

1945	Jan	Feb	Mar	Apr	May	June	July	Aug	Sept	Oct	Nov	Dec
TOTAL 5,435	857	721	758	693	710	701	602	208	65	94	26	–

Acknowledgments

The acknowledgments for a book are the way authors reveal who has guided and sustained them on the long road to completing the work. Mountaineers have scouts; writers have archivists. Athletes have coaches; writers have editors who shape their work. Our work would not have been possible without the invaluable efforts of certain individuals who deserve deep thanks for their help. Vivian Rogers-Price, research center director of the Roger A. Freeman Research Center at the National Museum of the Mighty Eighth Air Force in Pooler, Georgia, near Savannah, gave remarkable assistance in furnishing countless documents, memoirs, diaries, letters, military documents and reports, despite the coronavirus, which frustrated our research for some two years. When COVID-19 made it impossible for us to visit the museum, Dr. Rogers-Price arranged to search through and email reams of digitized records to us in New York in an extraordinary effort that went above and beyond all call of duty. Without her, our work would have been impossible. Debbie

ACKNOWLEDGMENTS

Seracini of the San Diego Air & Space Museum research library was our invaluable pathfinder through the voluminous archives there on Edgar Schmued and the design of the P-51 Mustang. Dorothy Alexander was a helpful guide at the National Archives and Records Administration in Washington, DC, for material on Bennett Meyers and corruption in the defense industry. Tammy Horton was a valuable conduit for documents at the U.S. Air Force Historical Research Agency in Montgomery, Alabama. When the coronavirus prevented our visit to the AFHRA, Ms. Horton provided digitized copies of important documents. Patrick Fahy at the Franklin D. Roosevelt Presidential Library and Museum led us through the papers of John Gilbert Winant.

Other individuals gave freely of their time and effort. General Michael P. C. Carns, USAF (ret.), was an ambassador to the Pentagon and officials there and granted both copious interview time and warm support. William Clifton of the U.S. Air Force historical office was most helpful and always answered our inquiries quickly and helpfully.

Louise Hitchcock Stephaich and the Hitchcock family were extraordinarily generous in furnishing recollections of Lieutenant Colonel Hitchcock, her father. Mrs. Stephaich's assistant, Perry Tuzcu, helped us immeasurably in providing annotated copies of Lieutenant Colonel Hitchcock's letters to his wife and recordings of interviews with many who knew him.

Others lent invaluable toil and care.

Our brilliant and visionary agent, James D. Hornfischer of Hornfischer Literary Management, LP, worked closely with us in shaping and polishing the proposal that led to our book; Jim's devotion to his authors and his tireless efforts on their behalf were remarkable. Tragically, Jim died in June 2021, one month before our book was completed. We miss his hand in the last mile of our voyage. Jim was a talented and prolific writer himself, the author of absorbing and highly successful books about the U.S. Navy in the Pacific.

Brent Howard, our editor at Dutton Caliber, was always a wise, studious, keen and patient editor who did an enormous amount to shape and hone the manuscript. Throughout our work he was both understanding and inspired. Grace Layer did important editing work on the project. Finally, our daughter, Margaret, never dimmed the smile of her encouragement or her support for her parents, and did yeoman labor typing drafts of the manuscript. And Thomas M. Stanback, Jr., formerly of the 20th Fighter Group, contributed in innumerable ways to the project. Tom sat for five interviews about the men of the 20th and answered countless questions; his stories of the lives and times of the pilots and crews of the 20th Fighter Group were the genesis of this work. We owe him all our thanks; in many ways this is Tom's book.

—New York, February 2022

Notes

ABBREVIATIONS:

SDASM: San Diego Air & Space Museum
NMMEAF: National Museum of the Mighty Eighth Air Force
NARA: National Archives and Records Administration
AFHRA: U.S. Air Force Historical Research Agency
FDRML: Franklin D. Roosevelt Presidential Museum and Library

Prologue: The Landmarks of a Nightmare

For more detail on the October Schweinfurt raid, see Bendiner, Elmer; *The Fall of Fortresses*; New York: G. P. Putnam's Sons, 1980.

2   **"the landmarks of a nightmare":** Ibid., p. 221.

Chapter 1: Hypnotized

For material on Edgar Schmued's early life and beginnings, see chapter 2 in Wagner, Ray; *Mustang Designer: Edgar Schmued and the P-51*; Washington, DC: Smithsonian Institution Press, 1990. For material on the origins of flight and the birth of aviation, see chapters 1 and 2 in Grant, R. G.; *Flight: The Complete History of Aviation*; New York: DK Publishing, 2017. For more on the Wright brothers, see McCullough, David; *The Wright Brothers*; New York: Simon & Schuster Paperbacks, 2015.

17   **"any aeronautical material":** Peace Treaty of Versailles, Articles 159–213, Military, Naval and Air Clauses, net.lib.byu.edu.

Chapter 2: *Flugzeug Konstructeur*

For more on the early years of the mature Edgar Schmued, see chapter 2 in Wagner, *Mustang Designer*. For information on the advent of early aviation and its golden age, see Grant, *Flight*. For Lindbergh's flight and the reaction to it, see Berg, A. Scott; *Lindbergh*; New York: G. P. Putnam's Sons, 1998. For the development of commercial aviation, see Yenne, Bill; *The American Aircraft Factory in World War II*; Minneapolis: Zenith Press, 2010.

19　**"I was living":** Wagner, p. 30.
19　**"so I could really utilize":** Ibid.
20　**"There is a particular flavor":** de Saint-Exupéry, Antoine, *Wind, Sand and Stars*, London: Heinemann, 1970, p. 160.
22　**"People behaved":** Berg, p. 170.
24　**"quality of work":** Wagner, p. 31.
25　**"clean, neat and exact":** Ibid., p. 30.
25　**"a welded steel tube":** Ibid., p. 33.
26　**"extrovert with an extraordinary":** *U.S. Air Services*, August 1939, p. 34.

Chapter 3: 100 Days

For the design and construction of the first P-51, see Wagner, *Mustang Designer*. For the significance of Schmued's design, see Anderson, John D., Jr.; *The Grand Designers: The Evolution of the Airplane in the 20th Century*; Cambridge, UK: Cambridge University Press, 2018.

30　**"Ed, do we want to build":** Wagner, p. 51.
31　**"Well, Dutch," Edgar replied:** Ibid.
31　**"Ed, I'm going":** Ibid.
32　**"I made many sketches":** Ibid.
32　**"Make it the fastest":** Ibid., pp. 51–52.
32　**"Meanwhile, the British":** Ibid., p. 53.
33　**"an Englishman had just":** Ibid.
34　**"Then, as you might say":** SDASM: interview with Edgar Schmued.
34　**"that the airplane took shape":** Ibid.
35　**"the Mustang materialized":** Rees, Ed, "A Tribute to Dutch Kindelberger," *Air Power Historian*, October 1962, pp. 197–206.
35　**"Sunday was different":** SDASM: interview with Schmued.
35　**"It was the greatest team":** Ibid.
36　**"The air likes that":** Wagner, p. 57.
36　**"I laid out the lines":** Ibid.
37　**"So God invented":** SDASM: Edward J. Horkey, letter to Ray Wagner, November 21, 1988.
38　**"[he] would drop around":** Wagner, p. 59.
38　**"When they saw the drawing":** Ibid., p. 60.

38 **"I told him let's not worry":** SDASM, interview with Schmued.

39 **"I should kick your ass":** SDASM, memo, recollections of Vern Tauscher.

39 **"When 102 days were over":** Wagner, p. 60.

39 **"The Allison . . . people told us":** Ibid., p. 64.

40 **"Don't come home":** Ibid.

Chapter 4: Death from Above

For background on the Battle of Britain, see Kennedy, Paul; *Engineers of Victory: The Problem Solvers Who Turned the Tide in the Second World War*; New York: Random House, 2013; Korda, Michael; *With Wings like Eagles: The Untold Story of the Battle of Britain*; New York: Harper Perennial, 2010; and Townsend, Peter; *Duel of Eagles*; Edison, NJ: Castle Books, 2003. For the development of radar, see Fisher, David E.; *A Race on the Edge of Time: Radar—The Decisive Weapon of World War II*; New York: McGraw-Hill, 1988.

46 **"There are none":** Fisher, David E., *A Summer Bright and Terrible: Winston Churchill, Lord Dowling, Radar, and the Impossible Triumph of the Battle of Britain*, Berkeley, Shoemaker & Hoard, 2005, p. 223.

46 **"Don't speak to me":** Townsend, p. 326.

46 **"bright and terrible":** Snow, C. P., *The Light and the Dark*, Cornwall, UK: House of Stratus, 1947, p. 302.

48 **"For Londoners, there are":** Panter-Downes, Molly, *London War Notes, 1939–1945*, New York: Farrar, Straus and Giroux, 1971, p. 98.

Chapter 5: A Turbine like a Typhoon

For the history of Rolls-Royce, see Pugh, Peter; *The Magic of a Name: The Rolls-Royce Story, Part One: The First 40 Years*; Cambridge, UK: Icon Books, 2000; and Winchester, Simon; *The Perfectionists: How Precision Engineers Created the Modern World*; New York: HarperCollins, 2018. For R. J. Mitchell and the design of the Spitfire, see Anderson, *The Grand Designers*. For the story of Stanley Hooker and the supercharger, see Harker, Ronald W.; *The Engines Were Rolls-Royce: An Informal History of That Famous Company*; New York: Macmillan Publishing Company, 1979.

52 **"impudent speech and a tiny waist":** "Saviour of the Spitfire," www.telegraph.co.uk.

Chapter 6: The Butcher Bird

For British acceptance and first production of the P-51A, see Wagner, *Mustang Designer*. For the arrival of the P-51A in the UK and use by the RAF, see Lowe, Malcolm V.; *North American P-51 Mustang*; Marlborough, UK: Crowood Press Ltd., 2009. For the appearance of the Fw 190, see Bader, Douglas; *Fight for the Sky: The Story of the Spitfire and the Hurricane*; Barnsley, UK: Pen and Sword Military, 2003. For Harker's flight in the P-51A, see Harker, *The Engines Were Rolls-Royce*; and Kennedy, *Engineers of Victory*.

58   **"The Focke-Wulf 190 certainly gave":** Simkin, John, "Focke Wulf 190A," https://spartacus-educational.com/2WWfocke190, September 1997 .htm.

Chapter 7: High Noon

For British reaction to Harker's flight, see Birch, David; *Rolls-Royce and the Mustang, Historical Series No. 9*; Derby, UK: Rolls-Royce Heritage Trust, 1987. For the British conversion of the P-51A to a Merlin engine, see Kennedy, *Engineers of Victory*; and Olson, Lynne; *Citizens of London: The Americans Who Stood with Britain in Her Darkest, Finest Hour*; New York: Random House, 2010. For Schmued's conversion, see Wagner, *Mustang Designer*.

61   **"The point which strikes me":** Birch, p. 10.
61   **"Full of excitement":** Harker, p. 70.
62   **"Churchillian in physique":** Ibid., p. xii.
63   **"With all the enthusiasm":** Ibid., p. 70.

Chapter 8: The Best Sport in the World

For the Japanese attack on Pearl Harbor, see Manchester, William, and Paul Reid; *The Last Lion: Winston Spencer Churchill, Defender of the Realm, 1940–1945*; New York: Little, Brown, 2012; and Winant, John G.; *A Letter from Grosvenor Square: An Account of a Stewardship*; London: Hodder and Stoughton, 1947. For the life and background of Lieutenant Colonel Tommy Hitchcock, see Aldrich, Nelson W., Jr.; *American Hero: The True Story of Tommy Hitchcock—Sports Star, War Hero, and Champion of the War-Winning P-51 Mustang*; Guilford, CT: Lyons Press, 1984.

72   **"I will call up the president":** Winant, p. 199.
72   **"It's quite true":** Manchester, p. 422.
73   **"The days of peace were over for us":** Winant, p. 200.
75   **"in speakeasies and private houses":** Aldrich, p. xxiii.
75   **"He would sit in the house":** Interview with Louise Hitchcock Stephaich, May 16, 2019.
75   **"I sat helpless and watched the ball":** Ballard, Sarah, "Polo Player Tommy Hitchcock Led a Life of Action from Beginning to End," *Sports Illustrated*, https://vault.si.com/vault/1986/11/03/polo-player-tommy-hitchcock -led-a-life-of-action-from-beginning-to-end.
76   **"a camel hair coat draped":** Madison, George, "Tommy Hitchcock: A Ten Goal American," personal collection, Louise Hitchcock Stephaich.
76   **"his own celebrity had gone so far":** Aldrich, p. 168.
76   **"wide, thick shoulders":** Ibid., p. 235.
76   **"gentleness of his voice":** Ibid.
78   **"I live for the flying":** Ibid., p. 221.
79   **"All I had to do":** Tommy Hitchcock, notes quoted in Aldrich, p. 91.
79   **"I was free":** Aldrich, p. 92.
80   **"from Sands Point":** Ibid., p. 223.
80   **"enjoyed a brief golden age":** Grant, p. 157.
80   **"spacious compartments, upholstered chairs":** Ibid.

81   **"Polo is exciting":** Olson, p. 251.
81   **"Socially, wartime Washington":** Aldrich, p. 235.

Chapter 9: A One-Man Crusade

For material on the later years of Lieutenant Colonel Tommy Hitchcock in London, see Aldrich, *American Hero*. For the campaign for the Mustang in England, see Birch, *Rolls-Royce and the Mustang*; Olson, *Citizens of London*; and Kennedy, *Engineers of Victory*. For more on the growth of the Eighth Air Force in England, see Parton, James; *Air Force Spoken Here: General Ira Eaker and the Command of the Air*; Bethesda, MD: Adler and Adler, 1986; and Coffey, Thomas M.; *Hap: The Story of the U.S. Air Force and the Man Who Built It, General "Hap" Arnold*; New York: Viking Press, 1982. For more on Chesley Peterson, see Haugland, Vern; *The Eagle Squadrons: Yanks in the RAF, 1940–1942*; New York: Ziff-Davis Flying Books, 1979.

83   **"It was now":** Richler, Mordecai, ed., *Writers on World War II: An Anthology*, New York: Vintage, 1993, p. 69.
84   **"London is dreary":** Tommy Hitchcock, letter to Margaret Mellon Hitchcock, May 1, 1942.
84   **"When I wake up here":** Aldrich, p. 243.
84   **"spent long hours poring over":** Ibid., p. 241.
85   **"It is estimated":** Ibid.
85   **"Yesterday I had a good flight":** Tommy Hitchcock, letter to Margaret Mellon Hitchcock, June 18, 1942.
86   **"a gentle, dreamy idealist":** Seib, Philip, *Broadcasts from the Blitz: How Edward R. Murrow Helped Lead America into War*, Washington, DC: Potomac Books, 2006, p. 121.
88   **"food, clerical help":** Parton, p. 142.
90   **"On Sundays, the Wrights would invite":** Coffey, p. 46.
92   **"They sowed the wind":** *Air Force* magazine, September 1, 2011.
92   **"like a club room":** Aldrich, p. 244.
93   **"could see first-hand":** Furse, Anthony, *Wilfrid Freeman: The Genius Behind Allied Survival and Air Supremacy, 1939 to 1945*, Staplehurst, UK: Spellmount, 1999, p. 229.
94   **"After he'd taken off":** Aldrich, p. 245.
94   **"The Mustang is one of the best":** Birch, pp. 37–38.
95   **"USAAF losses were four destroyed":** Kennedy, p.113.
96   **"It looks very much as if":** Tommy Hitchcock, letter to Margaret Mellon Hitchcock, September 16, 1942.
96   **"I hear you are heading for the States":** Haugland, p. 157.
97   **"was a beautifully handling plane":** Ibid.

Chapter 10: A Cross-Country Shuttle

For more information on Lieutenant Colonel Tommy Hitchcock's campaign for the P-51 Mustang in the U.S., see Aldrich, *American Hero*, and Olson, *Citizens of London*. For more on the Arsenal of Democracy, see Cooke, Alistair; *The American Home Front, 1941–1942*; New York: Atlantic Monthly Press, 2006.

100 **"I remember insisting":** Aldrich, p. 247.

100 **"The P-51 is not in the inventory":** Haugland, p. 158.

100 **"The P-51 is finished":** Ibid.

100 **"They didn't listen to me":** Ibid.

101 **"I am told by a young American friend":** FDRML: President Franklin D. Roosevelt, note to General Hap Arnold, November 10, 1942, Franklin D. Roosevelt Presidential Museum and Library.

101 **"sent a memo to General Echols":** Ludwig, Paul A., *The P-51 Mustang: Development of the Long-Range Escort Fighter*, Surrey, UK: Classic Publications, 2003, p. 131.

101 **"His hands were tied":** Olson, p. 262.

101 **"Hitchcock took on the job":** Aldrich, p. 247.

101 **"He appointed himself the ramrod":** Olson, p. 262.

102 **"spent all day talking airplanes":** Aldrich, p. 247.

102 **"There was a little excitement":** Tommy Hitchcock, letter to Margaret Mellon Hitchcock, December 15, 1942.

102 **"The Mustang still looks":** Ibid.

102 **"thousands . . . who stream[ed]":** Cooke, p. 139.

102 **"wings of bombers":** Ibid.

Chapter 11: Cash and Corruption

For more on delays in testing the P-51 and possible Army Air Forces corruption, see Ludwig, *The P-51 Mustang*. For competing priorities in the Army Air Forces, see Kennedy, *Engineers of Victory*. For the work of the Truman Committee, see McCullough, David, *Truman*, New York: Simon & Schuster, 1992.

105 **"The P-51 appears":** Ludwig, p. 74.

105 **"partly the frustration involved":** Edward R. Murrow, quoted in Nicholson, Nigel, ed., *Harold Nicholson, Diaries & Letters Volume II: The War Years, 1939–45*, New York: Atheneum, 1967, p. 226.

105 **"Sired by the English":** Kennedy, p. 124.

106 **"many Americans believed":** Ibid., p. 123.

106 **"the Allison-powered Mustang":** Aldrich, p. 247.

106 **"We are drawing freely":** Ibid., p. 248.

109 **"It's going on today":** Interview with General Michael P. C. Carns, August 10, 2019.

110 **"peppery little man":** NARA: Courson, George, "The Fantastic Story of General Meyers," testimony before the Truman Committee, November to December 1947, p. 1.

110 **"with Mr. Lamarre's knowledge":** Ibid., p. 23.

110 **"a cozy setup for the cozy war":** "Investigations: Rotten Apple," *Time*, December 1, 1947.

111 **"owned by some friends":** Courson, p. 4.

112 **"The experience . . . was an eye opener":** McCullough, *Truman*, p. 256.

112 **"Slightly built":** Ibid., p. 265.

112 **"travelled the length of the country"**: Ibid., p. 263.
113 **"practice as soon as he was in possession"**: Courson, p. 10.
113 **"stomped into the hearing"**: Ibid., p. 24.
113 **"disgraced his uniform"**: Ibid.
114 **"spark plug"**: NARA, "Activities of Maj. Gen. Bennett Mayers," testimony before the Truman Committee, November to December 1947.
117 **"the story of the P-51 came close"**: Kennedy, p. 124.

Chapter 12: Huddle in the Orange Groves

For more on the Casablanca conference, see Manchester, *The Last Lion*; and Macmillan, Harold; *The Blast of War, 1939–1945*; London: Macmillan, 1967. On Eaker's meeting with Churchill, see Parton, *Air Force Spoken Here*. On Hitchcock's return to England, see Aldrich, *American Hero*.

120 **"because Casablanca was well"**: Parton, p. 229.
120 **"at least three yards wide"**: Roberts, Andrew, *Masters and Commanders: How Four Titans Won the War in the West, 1941–1945*, New York: HarperCollins, 2009, p. 316.
120 **"almost every conceivable personality"**: Macmillan, Harold, *War Diaries: Politics and War in the Mediterranean, January 1943–May 1945*, New York: St. Martin's Press, 1984, p. 8.
121 **"you would see field marshals"**: Ibid., p. 9.
121 **"a curious mixture"**: Ibid., p. 8.
121 **"the scene [often] lit"**: Manchester, p. 626.
122 **"Mr. Churchill motioned me"**: Parton, p. 221.
123 **"the most lovely spot in the world"**: Manchester, p. 633.
124 **"dressed in velvet slippers"**: Ibid.
124 **"Seeing you and the children"**: Aldrich, p. 250.

Chapter 13: Picked Off like Geese

For more on the Ploeşti mission, see Ardery, Philip; *Bomber Pilot: A Memoir of World War II*; Lexington: University Press of Kentucky, 1978. For the psychological toll on bomber crews, see Miller, Donald L.; *Masters of the Air: America's Bomber Boys Who Fought the Air War Against Nazi Germany*; New York: Simon & Schuster Paperbacks, 2006. For Arnold's impatience with Eaker, see Parton, *Air Force Spoken Here*.

127 **"We were very close"**: Ardery, p. 103.
127 **"As their bombs were dropping"**: Ibid.
128 **"Many [planes] were so riddled"**: Ibid., p. 107.
132 **"Bomber bases were damn depressing places"**: Olson, p. 258.
134 **"Well, he sat down and started to eat grass"**: Interview with Mort Harris, November 11, 2019.
134 **"I didn't have the guts"**: Ibid.
134 **"Altogether, we started out for Berlin"**: John A. Miller, as quoted in Miller, p. 278.

135 **"It was exciting"**: Interview with J. Clifford Moos, October 30, 2019.

135 **"It was cold-blooded murder"**: Interview with Arnold Hague, April 29, 2002.

137 **"Under enormous pressure"**: Olson, p. 263.

137 **"totally indifferent to the fact"**: Boyne, Walter J., *Clash of Wings: World War II in the Air*, New York: Touchstone, 1994, p. 284.

137 **"They were true believers"**: Salisbury, Harrison E., *A Journey for Our Times*; New York: Carroll & Graf, pp. 194, 197.

138 **"My wire was sent to you"**: Parton, p. 272.

138 **"Arnold was terribly impatient"**: Ibid., p. 320.

139 **"a bouquet one day"**: Ibid., p. 319.

139 **"stoic strength and unfailing sense of humor"**: Ibid., p. 172.

141 **"Hap was having a hell of a time"**: Coffey, p. 321.

141 **"On this trip"**: Ibid., p. 322.

Chapter 14: Black Thursday

For background on the Schweinfurt raid, see Caidin, Martin; *Black Thursday: The Story of the Schweinfurt Raid*; New York: Dutton, 1960. For background on Robert Lovett, see Jordan, David M.; *Robert A. Lovett and the Development of American Air Power*; Jefferson, NC: McFarland & Company, Inc., 2019. For more on extending the P-51's range, see Parton, *Air Force Spoken Here*, and Coffey, *Hap*. On the American response to bomber losses, see Olson, *Citizens of London*.

143 **"You get the four engines going"**: Interview with Harris.

143 **"We were flying over an undercast"**: NMMEAF: recollections of Lieutenant Robert L. Hughes, 100th Bomb Group.

144 **"Once we were on our own"**: NMMEAF: recollections of anonymous bomber crewman, 94th Bomb Group, 333rd Bomb Squadron.

144 **"I had never seen"**: NMMEAF: memoir of Technical Sergeant George G. Roberts, radio operator, bomber *Cavalier*, 306th Bomb Group, 367th Bomb Squad.

145 **"They knew where we were"**: Interview with Harris.

145 **"I saw that little yellow flame go up"**: Ibid.

146 **"We just liked each other like brothers"**: Ibid.

147 **"Our plane shook from a 20-mm shell"**: NMMEAF: memoir of Technical Sergeant George G. Roberts.

147 **"One by one our gunners called in"**: NMMEAF: recollections of anonymous bomber crewman.

147 **"I'm hit, I'm hit"**: Ibid.

148 **"It was then time for my mission prayer"**: NMMEAF: memoir of Technical Sergeant George G. Roberts.

148 **"Limping back to the Channel"**: Ibid.

149 **"in mid-October the weather"**: Huston, Major General John W., ed., *American Airpower Comes of Age: General Henry H. "Hap" Arnold's World War II Diaries, Vol. 2*, Maxwell AFB, AL: Air University Press, 2002, p. 252.

150 **"Tall, thin, bald"**: Parton, p. 269.

150 **"the Luftwaffe, their country's aircraft":** Jordan, p. 30.

151 **"the immediate need":** Ibid., p. 76.

152 **"I pushed hard on Hap":** Ibid., p. 77.

152 **"Within this next six months":** Parton, p. 279.

152 **"High hopes are felt":** Ibid.

153 **"the wings would not be strong enough":** Coffey, p. 308.

153 **"Have a P-51 on the line":** Ibid.

153 **"didn't believe the plane":** Ibid.

153 **"Put the tanks in":** Ibid.

153 **"the water-laden P-51":** Ibid.

153 **"I ran out of gas":** Wagner, p. 114.

154 **"in his Washington ivory tower":** Olson, p. 263.

154 **"that their basic operational assumption":** Kennedy, p. 124.

Chapter 15: A Maverick Fighter Ace

For more material on Colonel Blakeslee in his early years and with the 4th Fighter Group, see Hall, Grover C., Jr.; *1,000 Destroyed: The Life and Times of the 4th Fighter Group*; Fallbrook, CA: Aero Publishers, 1978. For more on General Doolittle's fighter strategy, see Boyne, *Clash of Wings.*

158 **"the engines shut down":** Olds, Robin, with Christina Olds and Ed Rasimus, *Fighter Pilot: The Memoirs of Legandary Ace Robin Olds*, New York: St. Martin's Press, 2010, p. 95.

160 **"Two women?":** Haugland, p. 153.

160 **"probably the best":** Ibid., p. 144.

160 **"ruthless . . . relatively fearless":** Ibid.

160 **"In the air he was all business":** Blake, Steve, *The Pioneer Mustang Group: The 354th Fighter Group in World War II*, Atglen, PA: Schiffer Military History, 2008, p. 38.

161 **"It's the ship":** Hall, p. 135.

162 **"No sir, General":** Ibid., p. 136.

162 **"You couldn't tell enlisted men":** Ibid.

Chapter 16: The Jig Was Up

For more on the role of the 4th Fighter Group in the Berlin raid, see Hall, *1,000 Destroyed*; and Goodson, James A.; *Tumult in the Clouds: The Classic Story of the War in the Air*; New York: New American Library, 2004.

166 **"Well, you've seen":** Hall, p. 162.

166 **"Time to get up":** Ibid., p. 174.

167 **"Those 700 planes":** Interview with Lieutenant William R. MacClarence, September 22, 2019.

169 **"We had been over enemy territory":** NMMEAF: recollections of Lieutenant Charles A. Gilpin, Jr., 381st Bomb Group, 535th Bomb Squad.

170 **"I started closing":** NMMEAF: pilot's combat report of Lieutenant John T. Godfrey, 4th Fighter Group, March 6, 1944.

170 **"The Me 109 made three":** Ibid.

170 **"Immediately afterwards I saw":** NMMEAF: action report of Lieutenant Archie W. Chatterley, 4th Fighter Group, March 6, 1944.

170 **"doing such turns, never straightening":** Ibid.

171 **"I asked Major Mills":** Ibid.

171 **"When you were up there dogfighting":** Interview with Colonel Joseph A. Peterburs, USAF (ret.), October 22, 2021.

171 **"Major Mills edged the rest":** NMMEAF: action report of Lieutenant Archie W. Chatterley, March 6, 1944.

171 **"An Fw 190 came towards":** NMMEAF: confidential mission report of Lieutenant William A. Kazlawski, 379th Bomb Group, March 6, 1944.

171 **"[an] Fw 190 was making persistent":** NMMEAF: confidential mission report of Sergeant Dominick R. Giordano, 379th Bomb Group, March 6, 1944.

172 **"As we went down after him":** NMMEAF: action report of Lieutenant Howard N. Moulton, Jr., 4th Fighter Group, March 6, 1944.

173 **"I did a turn left and then right":** Ibid.

174 **"We got hit and all hell":** Interview with Peterburs, August 9, 2019.

175 **"B-17 going down":** NMMEAF: confidential mission report of Lieutenant Kenneth J. Duvall, 379th Bomb Group, March 6, 1944.

175 **"B-17 went down":** NMMEAF: confidential mission report of an anonymous bomber crewman, March 6, 1944.

175 **"Bombs were away on Berlin":** NMMEAF: lead navigator's narrative of Captain Andrew K. Dutch, March 6, 1944.

175 **"believed to be good":** Ibid.

175 **"It was sort of like":** Interview with Peterburs, October 27, 2021.

175 **"I held my fire":** NMMEAF: action report of Lieutenant Alexander Rafalovich, March 6, 1944.

176 **"I jumped three":** NMMEAF: action report of Lieutenant Nicholas Megura, March 6, 1944.

176 **"I raked the 3 Me 110s":** Ibid.

176 **"I climbed starboard":** Ibid.

177 **"One shell passed through":** NMMEAF: Captain Philip J. Field, letter, December 17, 1979.

177 **"It was the tail gunner":** Ibid.

179 **"I desire to commend":** NMMEAF: General Jesse F. Auton, commendation of the 4th Fighter Group, March 6, 1944.

179 **"On March 6, we finally made it":** NMMEAF: recollections of Lieutenant Charles A Gilpin, Jr.

180 **"The first time your bombers":** Olson, p. 268.

Chapter 17: Death of a Jockey

For more on Hitchcock in 1943 and on his death, see Aldrich, *American Hero*, and Olson, *Citizens of London*. For more on the P-51D and on Schmued's trip to England, see Wagner, *Mustang Designer*.

183 **"After he left":** Interview with Stephaich.

183 **"It was a beautiful little plane":** Interview with Thomas M. Stanback, Jr., April 14, 2017.

183 **"It was a whole different plane":** Interview with MacClarence.

184 **"just diving into the ground":** Aldrich, p. 267.

184 **"It was perhaps our most":** Ibid.

185 **"I was by myself":** Interview with Stephaich.

186 **"My mother was devastated":** Interview with Alexander M. Laughlin, July 22, 2019.

186 **"Don't forget":** Tommy Hitchcock, letter to Margaret Mellon Hitchcock, April 17, 1944.

187 **"This was the answer":** Wagner, p. 139.

187 **"American airplanes were heavier":** Ibid.

187 **"We used all these load factors":** Ibid, p. 140.

187 **"I went home":** Ibid.

188 **"His sense of duty":** Krock, Arthur; *New York Times*, April 21, 1944.

188 **"The man had no fear":** Daley, Arthur; *New York Times*, April 22, 1944.

Chapter 18: An Endless Roar Overhead

For more on Eisenhower's decision, see Kershaw, Alex; *The First Wave: The D-Day Warriors Who Led the Way to Victory in World War II*; New York: Dutton Caliber, 2019. For the role of the 4th Fighter Group on D-Day, see Hall, *1,000 Destroyed*, and Goodson, *Tumult in the Clouds*. On Operation Cobra, see Bradley, Omar N., and Clay Blair; *A General's Life: An Autobiography*; New York: Simon & Schuster, 1983.

190 **"OK," he said. "We'll go":** Kershaw, *The First Wave*.

190 **"I feel as though a sword":** Freidel, Frank, *Franklin D. Roosevelt: A Rendezvous with Destiny*, New York: Little, Brown, 1990, p. 526.

190 **"summoned to headquarters":** Hall, p. 292.

190 **"large black and white stripes":** Ibid.

191 **"I am prepared":** Ibid., p. 293.

192 **"Do you realize":** Manchester, p. 838.

192 **"I wake up at night":** Irving, p. 51.

192 **"If any blame":** Ibid., p. 150.

193 **"We had told Eisenhower":** Goodson, *Tumult in the Clouds*, p. 36.

193 **"We couldn't believe the ferocity":** Ibid., p. 37.

195 **"six foot high earthen walls":** Olson, p. 326.

195 **"I want it to be the biggest":** Atkinson, Rick, "Operation Cobra and the Breakout at Normandy," https://www.army.mil/article/42658/operation-cobra-and-the-breakout-at-normandy/.

196 **"a storm, or a machine":** Ernie Pyle quoted in Miller, p. 306.

196 **"Bomb run was 2 minutes in length":** NMMEAF: narrative report of Lieutenant Robert G. Littlejohn, 379th Bomb Group, July 25, 1944.

196 **"We bombed the primary visually":** NMMEAF: narrative report of Captain Rowland S. Williams, 379th Bomb Group, July 25, 1944.

197 **"The bombers came in":** normandyamericanheroes.com/blog-2/operation-cobra-the-9th-infantry-division.

197 **"This thing has busted wide open":** Atkinson, "Operation Cobra."
197 **"Things on our front":** Ibid.

Chapter 19: Big Brother, Little Friend

For more detail on the 20th Fighter Group, see Steiner, Edward J., et al., *King's Cliffe: The 20th Fighter Group and the 446th Air Service Group in the European Theatre of Operations*; Sheridan, WY: Sheridan Press, 2004.

203 **"You talk about a family":** Interview with Stanback.
203 **"There were nightly passes":** Ibid.
203 **"The word would be":** Ibid.
204 **"kind of laid-back":** Ibid.
205 **"He was a delightful guy":** Ibid.
205 **"I just love to fly little planes":** Colonel Robert P. Montgmery, quoted in interview with Stanback.
206 **"In the early 1940s":** Tuskegee Airmen publicity document.
207 **"We were told":** Interview with General Charles McGee, May 18, 2021.
207 **"If the plane had been a girl":** Ambrose, Stephen, *The Wild Blue: The Men and Boys Who Flew the B-24s over Germany, 1944–45*, New York: Simon & Schuster, 2001, p. 214.

Chapter 20: Oil Run

For the showdown on bomber offensive targets, see Miller, *Masters of the Air*, and Irving, *The War Between the Generals*. For Lieutenant Goodson's experience in the Pölitz raid, see Goodson, *Tumult in the Clouds*.

211 **"untidy and sometimes unshaven":** Irving, p. 71.
211 **"with a neat toothbrush mustache":** Ibid., p.73.
211 **"a small, mysterious man":** Grant, Rebecca, "The War on the Rails," *Air Force* magazine, August 1, 2007.
211 **"pipe-smoking, slim, urbane":** Irving, p. 73.
212 **"I am tired of dealing":** Irving, title page.
213 **"At times I'd say":** NMMEAF: diary of Lieutenant Veto A. Iavecchia.
214 **"Me 410s are attacking the plane behind us":** Ibid.
214 **"A few minutes later":** Ibid.
214 **"Suddenly our plane":** Ibid.
214 **"Suddenly I called 'Bombs away'":** Ibid.
214 **"Suddenly 7 Fw 190s":** Ibid.
215 **"I started after a Me 109":** NMMEAF: pilot's combat report of Lieutenant Donald R. Emerson, 4th Fighter Group, June 20, 1944.
215 **"There was a silver Mustang":** Ibid.
215 **"faster, more maneuverable":** Interview with Colonel Wallace E. Lowman, August 8, 2019.
215 **"We bounced about 15":** NMMEAF: pilot's combat report of Captain Otey M. Glass, Jr., 4th Fighter Group, June 20, 1944.
216 **"Lt. Gillette made the first pass":** NMMEAF: pilot's combat report of Lieutenant Charles H. Shilke, 4th Fighter Group, June 20, 1944.

217 **"Suddenly I was catching up":** Goodson, *Tumult in the Clouds*, p. 165.

217 **"At the same moment":** Ibid., p. 166.

218 **"I looked at my name":** Ibid.

218 **"Two more now":** Ibid.

219 **"We made our run":** NMMEAF: memoir of Colonel Chester B. Hackett, 389th Bomb Group.

219 **"Just after we dropped our bombs":** Ibid.

219 **"Dropping out of formation":** Ibid.

220 **"There was very little wind":** Ibid.

220 **"Lts. Monroe and Dickmeyer":** NMMEAF: action report, 4th Fighter Group, June 20, 1944.

221 **"and our pilots got five claims":** Ibid.

221 **"About two minutes past the target":** NMMEAF: diary of Lieutenant Veto A. Iavecchia.

221 **"I knew now":** Ibid.

221 **"I could see the coast":** Ibid.

222 **"I hurriedly asked":** Ibid.

222 **"Sweden," one of the men answered:** Ibid.

Chapter 21: A Perfect Show

For more material on the 4th Fighter Group's role in Operation Frantic, see Hall, *1,000 Destroyed*, and Goodson, *Tumult in the Clouds*.

226 **"I want you to land 68 aircraft":** Hall, p. 301.

227 **"Now look," he said:** Hall, p. 300.

227 **"This whole thing":** Hall, p. 302.

229 **"The end of a perfect show!":** Ibid., p. 304.

229 **"tougher than the trip over":** Goodson, *Tumult in the Clouds*, p. 79.

229 ***"Da, da. Schnapps?":*** Ibid., p. 306.

233 **"Hitler did make Europe":** Ibid., p. 80.

Chapter 22: Closing the Ring

For more on the failure to bomb Auschwitz and on indiscriminate bombing by the Americans, see Miller, *Masters of the Air*. On the Red Ball Express, see Irving, *The War Between the Generals*.

238 **"There were about 400 bombers":** Miller, p. 365.

239 **"The P-51, especially in regard":** Interview with Jeremy R. Kinney, National Air and Space Museum, Smithsonian Institution, May 19, 2021.

Chapter 23: Buzz Bombs and Doodlebugs

For more on the V-1s in London, see Manchester, *The Last Lion*. For details of Operation Aphrodite, see Miller, *Masters of the Air*.

245 **"Bombing results were observed":** NMMEAF: narrative report of Lieutenant Colonel Robert S. Kittel, 379th Bomb Group, August 4, 1944.

245 **"Our lead group got numerous"**: NMMEAF: narrative report of Captain John J. O'Connell, 379th Bomb Group, August 4, 1944.

245 **"shot up one He 111"**: AFHRA: intelligence report, 20th Fighter Group, August 4, 1944.

245 **"made a pass on Barth airfield"**: AFHRA: Ibid.

246 **"A flight of Me 262 twin engine jets"**: NMMEAF: memoir of Lieutenant Colonel Robert H. Kaurin, 381st Bomb Group.

247 **"I unhooked everything"**: NMMEAF: memoir of Captain Donald A. Reihmer, August 4, 1944.

247 **"The sea was rough"**: Ibid.

248 **"I was tossed and turned"**: Ibid.

248 **"While sailing I prayed"**: Ibid.

248 **"At about midnight"**: Ibid.

248 **"Are you English or American?"**: Ibid.

Chapter 24: Fighter Boys

For more detail on individual pilots of the 4th Fighter Group, see Goodson, *Tumult in the Clouds.*

252 **"I'm 19 years old"**: Interview with Peterburs, August 9, 2019.

252 **"You got the adrenaline"**: Interview with Lowman.

252 **"They had a swagger"**: Interview with MacClarence.

255 **"core of steel"**: Meyer, General John C., https://www.preddy-foundation .org/memorials/general-john-c-meyer/.

255 **"7. Always try to give"**: Preddy, George E., Jr., www.preddy-foundation .org.

256 **"The Kid was usually"**: Goodson, *Tumult in the Clouds,* p. 112.

256 **"I tried to draw him into discussions"**: Ibid.

257 **"It's most challenging"**: Interview with MacClarence.

257 **"You have the different types of aircraft"**: Interview with Peterburs, August 9, 2019.

Chapter 25: A Last Throw of the Dice

For more on the retirement from service of Colonel Blakeslee, see Hall, *1,000 Destroyed.* For more on the story of Joe Peterburs and Walter Schuck, see Peterburs, Joseph A.; *WW II Memories of a Mustang Pilot;* Joseph A. Peterburs. Self-published, 2007.

259 **"arguably the most beautiful"**: Boyne, p. 348.

260 **"It's more fun facing a squadron"**: "Hero of 500 Missions Flinches at Cameras," *New York Times,* September 12, 1944.

260 **"You are the symbol"**: Hall, p. 330.

260 **"When are you going"**: Ibid., p. 331.

262 **"The mission was one of"**: Interview with Peterburs, August 9, 2019.

263 **"I hit him"**: Ibid.

263 **"I felt a thud"**: Ibid.

263 **"I'm down to around 400 feet"**: Ibid.

Chapter 26: Twilight for the Reich

For more material on the last months of the war in the air for the 4th and 20th Fighter Groups, see MacKay, Ron; *20th Fighter Group*; Carrollton, TX: Squadron/ Signal Publications, 1995; and Davis, Larry; *4th Fighter Group in World War II*; Carrollton, TX: Squadron/Signal Publications, 2007.

265 **"A bitter, dense, freezing fog":** NMMEAF: memoir of Frank H. Lewis, December 24, 1944.
266 **"on December 23, as the Germans":** Ibid.
266 **"The field orders came clattering":** Ibid.
268 **"The outside temperature":** Ibid.
268 **"I watched the bombs all the way down":** Ibid.
268 **"I'll bet it was one of the greatest":** Ibid.
268 **"more full of hope":** Freidel, p. 572.
269 **"It's the biggest imposture":** *Encyclopedia Britannica*, "The World Wars, World War II," Chicago: Encyclopedia Britannica, 2007.
271 **"the loveliest city":** Miller, p. 428.

Epilogue: The Bells Toll

276 **"an imitation of a Blitz":** Panter-Downes, p. 373.
277 **"thousands of King George's subjects":** Ibid, p. 376.
277 **"When the crowd saw Churchill":** Ibid.
277 **"We formed up in the B-17s":** Interview with Stanback.
277 **"Kassel . . . just miles of rust":** Miller, p. 456.
278 **"The cities they had passed through":** Ibid.
280 **"one of the most gallant":** Aldrich, p. 271.
280 **"tangible evidence of Tommy's contribution":** Ibid.
281 **"Yup," he said:** Goodson, James A., *The Last of the Knights*, Canterbury, UK: Harrop Press, 1990, p. 140.

Bibliography

BOOKS

Agar, Herbert. *The Darkest Year: Britain Alone, June 1940–June 1941.* New York: Doubleday, 1973.

Aldrich, Nelson W., Jr. *American Hero: The True Story of Tommy Hitchcock—Sports Star, War Hero, and Champion of the War-Winning P-51 Mustang.* Guilford, CT: Lyons Press, 1984.

Ambrose, Stephen E. *Citizen Soldiers: The U.S. Army from the Normandy Beaches to the Bulge to the Surrender of Germany, June 7, 1944–May 7, 1945.* New York: Simon & Schuster, 1997.

———. *D-Day, June 6, 1944: The Climactic Battle of World War II.* New York: Simon & Schuster, 1994.

———. *The Supreme Commander: The War Years of Dwight D. Eisenhower.* New York: Doubleday, 1970; Anchor Books, 2012.

———. *The Wild Blue: The Men and Boys Who Flew the B-24s over Germany, 1944–45.* New York: Simon & Schuster, 2001.

Anderson, John D., Jr. *The Grand Designers: The Evolution of the Airplane in the 20th Century.* Cambridge, UK: Cambridge University Press, 2018.

Angelucci, Enzo, and Peter Bowers. *The American Fighter: The Definitive Guide to American Fighter Aircraft from 1917 to the Present.* New York: Orion Books, 1985.

Ardery, Philip. *Bomber Pilot: A Memoir of World War II.* Lexington: University Press of Kentucky, 1978.

Arnold, Henry H. *Global Mission.* New York: Harper & Brothers, 1949.

Atkinson, Rick. *The Guns at Last Light: The War in Western Europe, 1944–1945.* New York: Picador, 2013.

Bader, Douglas. *Fight for the Sky: The Story of the Spitfire and the Hurricane.* Barnsley, UK: Pen and Sword Military, 1973.

Baumbach, Werner. *Broken Swastika: The Defeat of the Luftwaffe.* London: Robert Hale, 1960.

Bendiner, Elmer. *The Fall of Fortresses.* New York: G. P. Putnam's Sons, 1980.

Berg, A. Scott. *Lindbergh.* New York: G. P. Putnam's Sons, 1998.

Billington, David P., and David P. Billington, Jr. *Power, Speed, and Form: Engineers and the Making of the Twentieth Century.* Princeton, NJ: Princeton University Press, 2006.

Birch, David. *Rolls-Royce and the Mustang, Historical Series No. 9.* Derby, UK: Rolls-Royce Heritage Trust, 1987.

Bishop, Patrick. *Bomber Boys: Fighting Back, 1940–1945.* London: Harper Perennial, 2007.

Blake, Steve. *The Pioneer Mustang Group: The 354th Fighter Group in World War II.* Atglen, PA: Schiffer Military History, 2008.

Bodie, Warren M. *The Lockheed P-38 Lightning: The Definitive Story of Lockheed's P-38 Fighter.* Hiawassee, GA: Widewing Publications, 1991.

Bowman, Martin. *Fighter Bases of World War II: 8th Air Force USAAF, 1943–45.* Barnsley, UK: Pen & Sword Aviation, 2009.

Boyne, Walter J. *Clash of Wings: World War II in the Air.* New York: Touchstone, 1994.

Bradley, Omar N., and Clay Blair. *A General's Life: An Autobiography.* New York: Simon & Schuster, 1983.

Brinkley, David. *Washington Goes to War.* New York: Ballantine Books, 1996.

Bull, Andy. *Speed Kings: The 1932 Winter Olympics and the Fastest Men in the World.* New York: Avery, 2015.

Caidin, Martin. *Black Thursday: The Story of the Schweinfurt Raid.* New York: Dutton, 1960.

Caine, Philip D. *American Pilots in the R.A.F.: The World War II Eagle Squadrons.* Washington, DC: Brassey's (U.S.), 1993.

Clapper, Olive Ewing. *Washington Tapestry.* New York: Whittlesey House, 1946.

Coffey, Thomas M. *Hap: The Story of the U.S. Air Force and the Man Who Built It, General Henry "Hap" Arnold.* New York: Viking Press, 1982.

Colville, John. *Footprints in Time: Memories.* Salisbury, UK: Michael Russell, 1984.

———. *The Fringes of Power: 10 Downing Street Diaries, 1939–1955.* New York: W. W. Norton, 1986.

———. *Winston Churchill and His Inner Circle.* New York: Wyndham Books, 1981.

Conant, Jennet. *The Irregulars: Roald Dahl and the British Spy Ring in Wartime Washington.* New York: Simon & Schuster Paperbacks, 2008.

Cooke, Alistair. *The American Home Front, 1941–1942.* New York: Atlantic Monthly Press, 2006.

Copp, DeWitt S. *A Few Great Captains: The Men and Events That Shaped the Development of U.S. Air Power.* McLean, VA: EPM Publications, 1980.

———. *Forged in Fire.* New York: Doubleday, 1982.

Craven, Wesley F., and James L. Cate, eds. *The Army Air Forces in World War II, Vol. II: Europe—Torch to Pointblank, August 1942 to December 1943.* Chicago: University of Chicago Press, 1949.

BIBLIOGRAPHY

————. *The Army Air Forces in World War II, Vol. III: Europe—Argument to V-E Day, January 1944 to May 1945*. Chicago: University of Chicago Press, 1951.

Daso, Dik Alan. *Hap Arnold and the Evolution of American Airpower*. Washington, DC: Smithsonian Institution Press, 2000.

Davis, Kenneth S. *FDR: The War President, 1940–1943*. New York: Random House, 2000.

Davis, Larry. *4th Fighter Group in World War II*. Carrollton, TX: Squadron/Signal Publications, 2007.

de Saint-Exupéry, Antoine. *Wind, Sand and Stars*. Translated by Lewis Galantière. New York: Harbrace Paperbound Library, 1967.

Eiffel, Gustave. *Recherches expérimentales sur la résistance de l'air executées à la tour Eiffel (Ed. 1907)*. Paris: L. Maretheux, 1907.

Ethell, Jeffrey. *Mustang: A Documentary History of the P-51*. New York: Jane's Publishing, 1981.

Fisher, David E. *A Race on the Edge of Time: Radar—The Decisive Weapon of World War II*. New York: McGraw-Hill, 1988.

————. *A Summer Bright and Terrible: Winston Churchill, Lord Dowding, Radar, and the Impossible Triumph of the Battle of Britain*. Berkeley: Shoemaker & Hoard, 2005.

Folly, Martin. *The United States and World War II: The Awakening Giant*. Edinburgh: Edinburgh University Press, 2002.

Fortier, Norman J. *An Ace of the Eighth: An American Fighter Pilot's Air War in Europe*. New York: Presidio Press, 2003.

Freeman, Roger A. *The Mighty Eighth: A History of the Units, Men and Machines of the U.S. Eighth Air Force*. London: Cassell, 2000.

————. *Mustang at War*. New York: Doubleday, 1974.

Freidel, Frank. *Franklin D. Roosevelt: A Rendezvous with Destiny*. New York: Little, Brown, 1990.

Fry, Garry L., and Jeffrey L. Ethell. *Escort to Berlin: The 4th Fighter Group in World War II*. New York: Arco, 1980.

Furse, Anthony. *Wilfrid Freeman: The Genius Behind Allied Survival and Air Supremacy, 1939 to 1945*. Staplehurst, UK: Spellmount, 1999.

Gilbert, Martin. *Churchill: A Life*. New York: Henry Holt, 1991.

————. *The Churchill Documents, Vol. 17: Testing Times, 1942*. Hillsdale, MI: Hillsdale College Press, 2014.

————. *Winston S. Churchill, Vol. VI: Finest Hour, 1939–1941*. Boston: Houghton, Mifflin, 1983.

Goodson, James A. *The Last of the Knights*. Canterbury, UK: Harrop Press, 1990.

————. *Tumult in the Clouds: The Classic Story of the War in the Air*. New York: New American Library, 2004.

Granger, Byrd Howell. *On Final Approach: The Women Airforce Service Pilots of World War II*. Scottsdale, AZ: Falconer, 1991.

Grant, R. G. *Flight: The Complete History of Aviation*. New York: DK Publishing, 2017.

Gruenhagen, Robert W. *Mustang: The Story of the P-51 Fighter*. New York: Arco Publishing, 1976.

Hall, Grover C., Jr. *1,000 Destroyed: The Life and Times of the 4th Fighter Group*. Fallbrook, CA: Aero Publishers, 1978.

BIBLIOGRAPHY

Hammel, Eric. *The Road to Big Week: The Struggle for Daylight Supremacy over Western Europe, July 1942–February 1944.* Pacifica, CA: Pacifica Military History, 2009.

Harker, Ronald W. *The Engines Were Rolls-Royce: An Informal History of That Famous Company.* New York: Macmillan, 1979.

Hastings, Max. *Winston's War: Churchill, 1940–1945.* New York: Alfred A. Knopf, 2009.

Haugland, Vern. *The Eagle Squadrons: Yanks in the RAF, 1940–1942.* New York: Ziff-Davis Flying Books, 1979.

———. *The Eagles' War: The Saga of the Eagle Squadron Pilots, 1940–1945.* New York: Jason Aronson, 1982.

Herman, Arthur. *Freedom's Forge: How American Business Produced Victory in World War II.* New York: Random House Trade Paperbacks, 2012.

Holley, Irving Brinton, Jr. *Buying Aircraft: Matériel Procurement for the Army Air Forces; United States Army in World War II; Special Studies.* Washington, DC: Center of Military History, 1989.

Hooker, Sir Stanley. *Not Much of an Engineer: An Autobiography.* Ramsbury, UK: Airlife, 2002.

Howard, James H. *Roar of the Tiger.* New York: Pocket Books, 1991.

Huston, Major General John W., ed. *American Airpower Comes of Age: General Henry H. "Hap" Arnold's World War II Diaries, Vols. 1 and 2.* Maxwell AFB, AL: Air University Press, 2002.

Ingersoll, Ralph. *Top Secret.* New York: Harcourt, Brace, 1946.

Irving, David. *The War Between the Generals.* New York: Congdon and Lattès, 1981.

Isaacson, Walter, and Evan Thomas. *The Wise Men: Six Friends and the World They Made.* New York: Simon & Schuster Paperbacks, 2012.

Jackson, Robert. *Warplanes of World War II.* London: Amber Books, 2018.

Jordan, David M. *Robert A. Lovett and the Development of American Air Power.* Jefferson, NC: McFarland, 2019.

Kelsey, Benjamin S. *The Dragon's Teeth? The Creation of United States Air Power for World War II.* Washington, DC: Smithsonian Institution Press, 1982.

Kennedy, Paul. *Engineers of Victory: The Problem Solvers Who Turned the Tide in the Second World War.* New York: Random House, 2013.

Kershaw, Alex. *The Few: The American "Knights of the Air" Who Risked Everything to Save Britain in the Summer of 1940.* Philadelphia: Da Capo Press, 2006.

———. *The First Wave: The D-Day Warriors Who Led the Way to Victory in World War II.* New York: Dutton Caliber, 2019.

Korda, Michael. *With Wings like Eagles: The Untold Story of the Battle of Britain.* New York: Harper Perennial, 2010.

Larrabee, Eric. *Commander in Chief: Franklin Delano Roosevelt, His Lieutenants and Their War.* New York: Harper & Row, 1987.

Lowe, Malcolm V. *North American P-51 Mustang.* Marlborough, UK: Crowood Press, 2009.

Ludwig, Paul A. *P-51 Mustang: Development of the Long-Range Escort Fighter.* Surrey, UK: Classic Publications, 2003.

MacKay, Ron. *20th Fighter Group.* Carrollton, TX: Squadron/Signal Publications, 1995.

Macmillan, Harold. *The Blast of War, 1939–1945.* London: Macmillan, 1967.

———. *War Diaries: Politics and War in the Mediterranean, January 1943–May 1945.* New York: St. Martin's Press, 1984.

Manchester, William, and Paul Reid. *The Last Lion: Winston Spencer Churchill, Defender of the Realm, 1940–1945.* New York: Little, Brown, 2012.

McCrary, John R., and David E. Sherman. *First of the Many: A Journal of Action with the Men of the Eighth Air Force.* London: Robson, 1981.

McCullough, David. *Truman.* New York: Simon & Schuster, 1992.

———. *The Wright Brothers.* New York: Simon & Schuster Paperbacks, 2015.

McFarland, Stephen L., and Wesley P. Newton. *To Command the Sky: The Battle for Air Superiority over Germany, 1942–1944.* Washington, DC: Smithsonian Institution Press, 1991.

Meacham, Jon. *Franklin and Winston: An Intimate Portrait of an Epic Friendship.* New York: Random House Trade Paperbacks, 2004.

Miller, Arthur. *All My Sons: A Drama in Three Acts.* New York: Dramatists Play Service, 1974.

Miller, Donald L. *Masters of the Air: America's Bomber Boys Who Fought the Air War Against Nazi Germany.* New York: Simon & Schuster Paperbacks, 2006.

Mundy, Liza. *Code Girls: The Untold Story of the American Women Code Breakers in World War II.* New York: Hachette Book Group, 2017.

Murray, Williamson. *Strategy for Defeat: The Luftwaffe, 1933–1945.* Maxwell AFB, AL: Air University Press, 1983.

Nichol, John. *Spitfire.* London: Simon & Schuster, 2019.

Nicholson, Nigel, ed. *Harold Nicholson, Diaries & Letters, Vol. II: The War Years, 1939–45.* New York: Atheneum, 1967.

Olds, Robin, with Christina Olds and Ed Rasimus. *Fighter Pilot: The Memoirs of Legendary Ace Robin Olds.* New York: St. Martin's Press, 2010.

O'Leary, Michael. *Building the P-51 Mustang: The Story of Manufacturing North American's Legendary World War II Fighter in Original Photos.* North Branch, MN: Specialty Press, 2011.

Olson, Lynne. *Citizens of London: The Americans Who Stood with Britain in Its Darkest, Finest Hour.* New York: Random House, 2010.

Pace, Steve. *Mustang: Thoroughbred Stallion of the Air.* Stroud, UK: Fonthill Media, 2012.

Panter-Downes, Mollie. *London War Notes, 1939–1945.* New York: Farrar, Straus and Giroux, 1971.

Parton, James. *Air Force Spoken Here: General Ira Eaker and the Command of the Air.* Bethesda, MD: Adler and Adler, 1986.

Peterburs, Joseph A. *The Autobiography of Joseph Anthony Peterburs, Colonel, USAF, Retired.* Self-published, 2013.

———. *World War II Memories of a Mustang Pilot.* Self-Published, 2007.

Price, Alfred. *Combat Development in World War II: Fighter Aircraft.* London: Arms and Armour Press, 1989.

———. *The Spitfire Story.* London: Arms and Armour Press, 1993.

Pugh, Peter. *The Magic of a Name: The Rolls-Royce Story, Part One: The First 40 Years.* Cambridge, UK: Icon, 2000.

———. *The Magic of a Name: The Rolls-Royce Story, Part Three: A Family of Engines.* Cambridge, UK: Icon, 2002.

Richler, Mordecai, ed. *Writers on World War II: An Anthology*. New York: Vintage, 1993.

Rickman, Sarah Byrn. *WASP of the Ferry Command: Women Pilots, Uncommon Deeds*. Denton, TX: University of North Texas Press, 2016.

Roberts, Andrew. *Masters and Commanders: How Four Titans Won the War in the West, 1941–1945*. New York: HarperCollins, 2009.

Rostow, W. W. *Pre-Invasion Bombing Strategy: General Eisenhower's Decision of March 25, 1944*. Austin: University of Texas Press, 1981.

Salisbury, Harrison E. *A Journey for Our Times*. New York: Carroll & Graf, 1984.

Sebald, W. G. *On the Natural History of Destruction*. Translated by Anthea Bell. London: Penguin, 2004.

Seib, Philip. *Broadcasts from the Blitz: How Edward R. Murrow Helped Lead America into War*. Washington, DC: Potomac, 2006.

Slessor, Sir John. *The Central Blue: Recollections and Reflections*. London: Cassell, 1956.

Snow, C. P. *The Light and the Dark*. Cornwall, UK: House of Stratus, 1947.

Speer, Frank E. *The Debden Warbirds: The 4th Fighter Group in World War II*. Atglen, PA: Schiffer, 1999.

Stacks, John F. *Scotty: James B. Reston and the Rise and Fall of American Journalism*. Boston: Little, Brown, 2003.

Stargardt, Nicholas. *The German War: A Nation Under Arms, 1939–1945*. New York: Basic Books, 2015.

Steiner, Edward J., Henry W. Howard, Jr., Ralph C. Hinds, Jr., and James W. Gatchell, eds. *King's Cliffe: The 20th Fighter Group and the 446th Air Service Group in the European Theatre of Operations*. N.p.: Sheridan Press, 2004.

Stiles, Bert. *Serenade to the Big Bird: A True Account of Life and Death from Inside the Cockpit*. New York: W. W. Norton, 1952.

Todman, Daniel. *Britain's War: Into Battle, 1937–1941*. New York: Oxford University Press, 2016.

Townsend, Peter. *Duel of Eagles*. Edison, NJ: Castle, 2003.

Turner, Richard E. *Big Friend, Little Friend: Memoirs of a World War II Fighter Pilot*. Mesa, AZ: Champlin Fighter Museum Press, 1983.

Wagner, Ray. *Mustang Designer: Edgar Schmued and the P-51*. Washington, DC: Smithsonian Institution Press, 1990

Wellum, Geoffrey. *First Light*. London: Penguin, 2009.

White, David Fairbank. *Bitter Ocean: The Battle of the Atlantic, 1939–1945*. New York: Simon & Schuster, 2006.

Whitney, Daniel D. *Vee's for Victory! The Story of the Allison V-1710 Aircraft Engine, 1929–1948*. Atglen, PA: Schiffer Military History, 1998.

Wilson, Thomas. *Churchill and the Prof*. London: Cassell, 1995.

Winant, John G. *A Letter from Grosvenor Square: An Account of a Stewardship*. London: Hodder and Stoughton, 1947.

Winchester, Simon. *The Perfectionists: How Precision Engineers Created the Modern World*. New York: HarperCollins, 2018.

Yenne, Bill. *The American Aircraft Factory in World War II*. Minneapolis: Zenith Press, 2010.

Ziegler, Philip. *London at War, 1939–1945*. London: Pimlico, 2002.

BIBLIOGRAPHY

ARCHIVAL SOURCES

National Museum of the Mighty Eighth Air Force

- Documents pertaining to the October 14, 1943, U.S. bombing raid on Schweinfurt, Germany.
- Documents pertaining to the March 6, 1944, U.S. bombing raid on Berlin, Germany.
- Documents pertaining to the June 20, 1944, U.S. bombing raid on Pölitz, Germany.
- Documents pertaining to the bombing in Operation Cobra, July 24–25, 1944.
- Documents pertaining to the August 4, 1944, U.S. bombing raid on Peenemünde, Germany.
- Documents pertaining to aerial combat in the Battle of the Bulge, December 1944–January 1945, and Operation Varsity, March 24, 1945.
- Documents pertaining to operations of Eighth Air Force, U.S. Army Air Forces, October 1943–May 1945.

San Diego Air & Space Museum

- Audio and written documents pertaining to Edgar Schmued.
- Documents pertaining to the development of the P-51 Mustang.

U.S. Air Force Historical Research Agency

- Documents pertaining to operations of the 20th Fighter Group, 1944.

National Archives and Records Administration

- Documents pertaining to hearings of the Senate Special Committee to Investigate the National Defense Program.

Franklin D. Roosevelt Presidential Library and Museum

- Papers of President Franklin D. Roosevelt, John Gilbert Winant, Lieutenant Colonel Tommy Hitchcock, Harry L. Hopkins, Bernard Bellush.

Letters

- Letters of Tommy Hitchcock and Margaret Mellon Hitchcock, May 1, 1942, to April 17, 1944.

INTERVIEWS

- Louise Hitchcock Stephaich, May 16, 2019
- Alexander M. Laughlin, July 22, 2019
- Colonel Wallace E. Lowman, USAF (ret.), August 8, 2019
- Colonel Joseph A. Peterburs, USAF (ret.), August 9, 2019; October 22, 2021; October 27, 2021
- General Michael P. C. Carns, USAF (ret.), August 10, 2019
- Thomas M. Stanback, Jr., August 25, 2019; November 24, 2019; March 27, 2021; March 29, 2021; April 4, 2021

BIBLIOGRAPHY

- William R. MacClarence, September 22, 2019
- Lieutenant J. Clifford Moos, October 30, 2019
- Mort Harris, November 11, 2019
- Dr. John I. Dintenfass, January 22, 2021
- Jeremy Kinney, May 19, 2021
- General Charles McGee, May 18, 2021
- Interviews by James Claggett and Nelson W. Aldrich, Jr., for Aldrich, *American Hero*
- Arnold Hague, April 29, 2002

WEBSITES (PRINCIPAL)

- acepilots.com
- afhra.af.mil
- army.mil/article/42658
- aviation-history.com
- bbc.co.uk
- boeing.com
- britannica.com
- history.com
- historylearningsite.co.uk
- historyplace.com
- homeofheroes.com
- militaryfactory.com
- mustangsmustangs.com
- net.lib.byu.edu
- newworldencyclopedia.org
- normandyamericanheroes.com
- preddy-foundation.org
- spartacus-education.com
- telegraph.co.uk/history/battle-of-Britain
- wikipedia.org

MAPS

- Map p. 89, "U.S. Eighth Air Force Installations in Britain, 1944," by Erichsen Group, with certain data from Jane's Publishing Co., Ltd.
- Map p. 130, "The Race for Range: Comparative Range of Allied Fighter Aircraft," by Erichsen Group, with certain data from Kennedy, *Engineers of Victory*.
- Map p. 224, "Selected Targets in the European Air War, 1943–45," by Erichsen Group.

Index

INDEX

INDEX

INDEX

About the Authors

David Fairbank White studied history at Harvard and worked as a reporter for the *New York Times*. He has written for national magazines including *Fortune, New York, Parade* and *Reader's Digest.* **Margaret Stanback White** received her BA from Harvard and did graduate work in public health at Yale. Following a long career as a research scientist, she began to freelance as a researcher and editor on nonfiction book projects about scientific breakthroughs and how they have changed the course of history.